CORK in Architecture, Design, Fashion, and Art

Imprint

Published 2019 by
GINGKO PRESS

Gingko Press Verlags GmbH
Schulterblatt 58
D-20357 Hamburg
Germany
Tel: +49 (0)40-291425
Fax: +49 (0)40-291055
Email: gingkopress@t-online.de

Gingko Press Inc
2332 Fourth Street, Suite E
Berkeley, CA 94710
USA
Tel: (510) 898 1195
Fax: (510) 898 1196
Email: books@gingkopress.com
www.gingkopress.com

ISBN: 978-3-943330-32-8

English translation: Michelle Bailat-Jones © 2019 Gingko Press Verlags GmbH
Copyediting: Rebecca Rider
Typesetting: Annalena Weber

Printed in Slovenia

Original title: LE LIÈGE
Originally published in French by Editions Gallimard, collection Alternatives, Paris, 2017

For the original French edition:
Graphic Design and Production: Line Célo
Assisted by Anna Petitcollot

CORK in Architecture, Design, Fashion, and Art

\

Chloé Genevaux
Guillaume Bounoure

GINGKO PRESS

PREFACE

The challenges for humans in the twenty-first century are to develop industries based on renewable resources, to decrease pollution, and to better manage our ecosystems.

This means that the products of the future must be imbued with this new ethos and ethical practices at the local level that include transparency and a green lifecycle.

Tomorrow's innovative materials will be like future generations: more frugal, more durable, and more genetically diverse.

The sustainable use of natural resources associated with both high-tech and traditional know-how is potentially a future path for all of us. For example, small units of local production that rely on locally available resources will become an integral part of a more cohesive land-development philosophy.

Innovation will create pioneers out of individuals and companies in this new economy. Technological advances and a greater distribution of knowledge will enable us to reinvent the objects of our daily lives.

So, what are the most meaningful materials in this context, and how do we use them?

Cork is a fantastic material for design: as an insulator, as a type of plant-based leather, or as a composite material for use in high-tech applications. It is beautiful and multifaceted, and it has an exemplary lifecycle and durability. Considering its qualities, cork is primed to become a component of choice for objects of the future, such as vehicles, furniture, lights, clothing, and fashion accessories, and it is also a highly prized material for many artists who use it both for its aesthetic beauty and its functional qualities.

In these pages, we feature the work of artists and designers who are using cork. You will discover numerous useful references and practical information about the use of this material in all its different forms: from production to transformation to finished product. This book is designed to be a source of inspiration—an occasion to discover a high-quality raw material produced by a living organism, the cork oak.

/ Strips of cork wood from which bottle corks have been extracted
© APCOR

/ Tumiki + building blocks for children designed by Aya Koike for Cork+
©Aya Koike

FROM TREE TO OBJECT

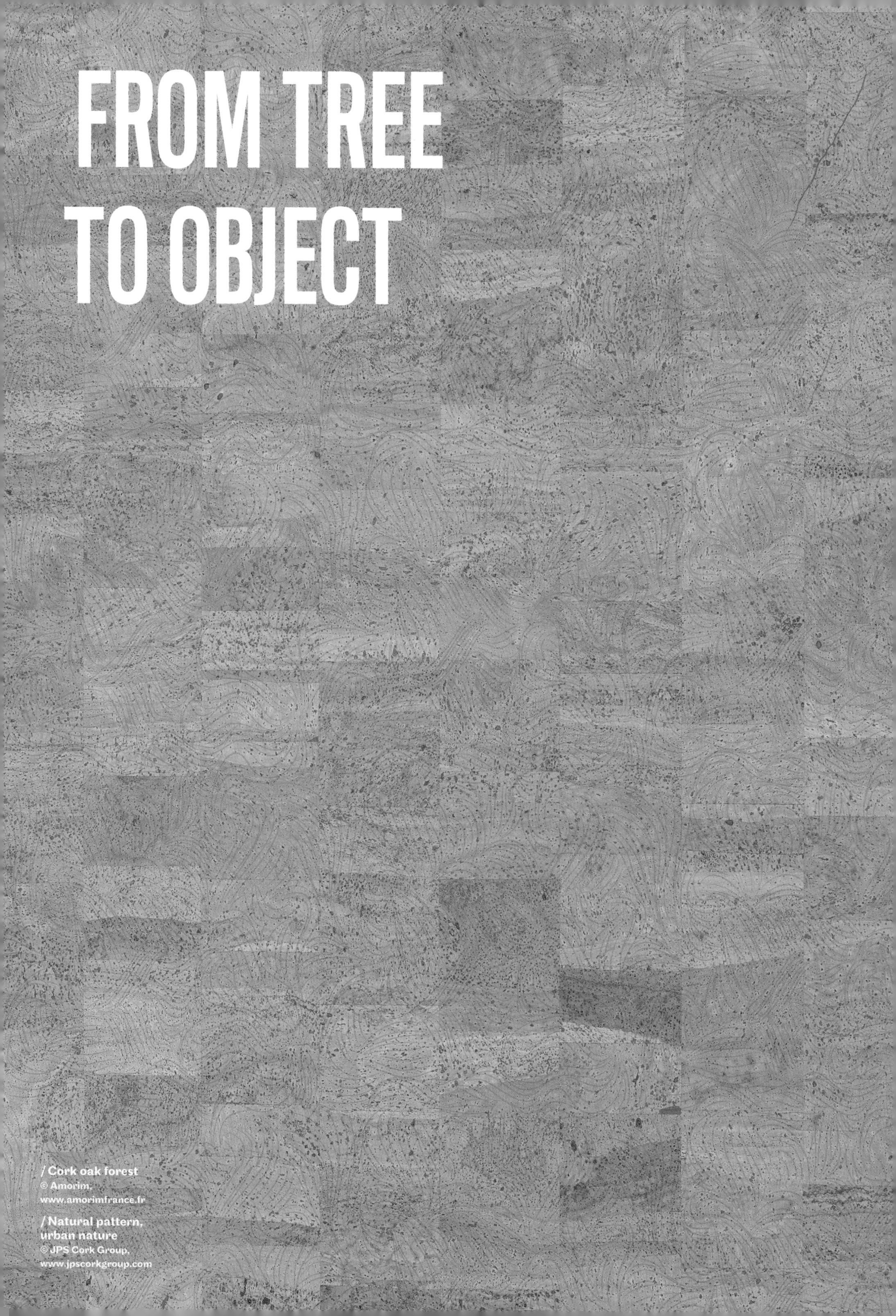

/ Cork oak forest
© Amorim,
www.amorimfrance.fr

/ Natural pattern,
urban nature
© JPS Cork Group,
www.jpscorkgroup.com

THE TREE AND ITS ECOSYSTEM

The cork oak is a species found around the entire Mediterranean. Its origins date back to the formation of the Mediterranean basin approximately sixty million years ago. Its thick bark protects it from the cold, which is why it has survived through several ice ages and is with us today. Its lifespan ranges from 200 to 800 years, and it reaches an average of 50 feet (15 m) high, although the largest known specimen was recorded at 141 feet (43 m).

The cork oak, whose scientific name is *Quercus suber*, belongs to the *Fagaceae* family. *Quercus* means oak in Latin, whereas *suber* means cork. Cork oak forests play an important role in regulating the lifecycle of water and protecting soils—they are an especially important protection against desertification near the Sahara and the Sahel—and at the same time, they also help conserve rich biodiversity. They are also able to store carbon over a long period of time, which is a highly sought-after quality of cork in terms of its production.

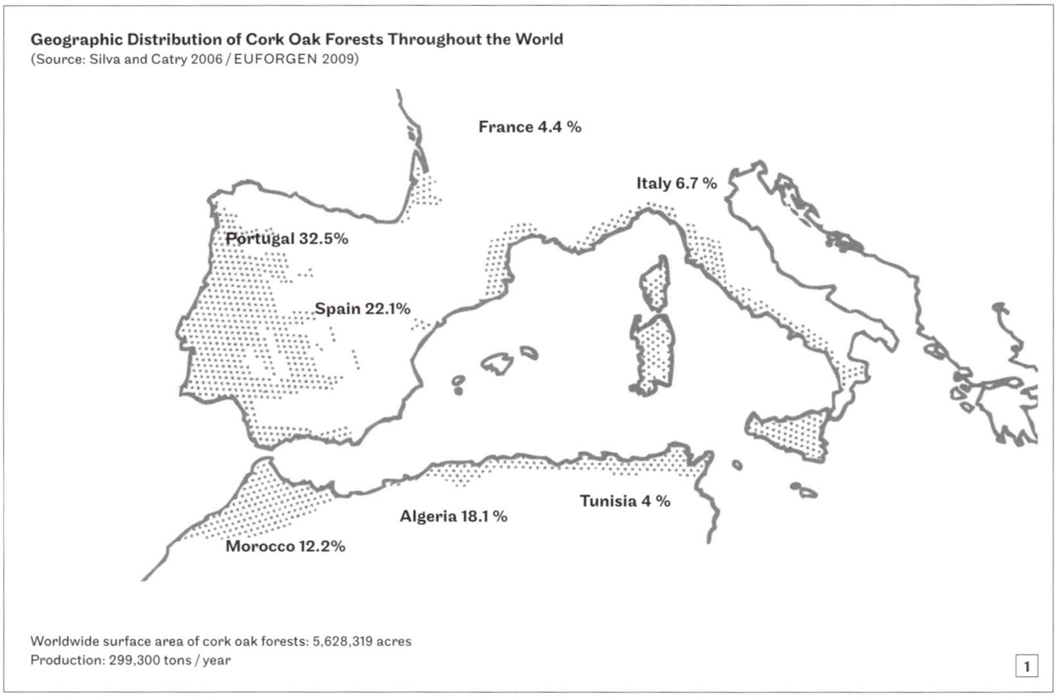

/
Area of Geographic Distribution

1 / Geographic distribution of cork oak forests throughout the world. The world's production of cork is divided between the seven countries in which its growth conditions are best: Portugal, Spain, Algeria, Morocco, Italy, Tunisia, and France. In order to grow well, the cork oak needs noncalcareous soils that are rich in silica and have a pH between 4.8 and 7.0; they also need annual rainfall amounts of between 15 and 30 inches (400 and 800 mm), mild annual temperatures of between 59 and 62° (12 to 19° C), and an altitude lower than 1640 feet (500 m).
© Mauricio O Brien / Corklab, Elisalva

2 / Young cork oak plant
© Amorim

3 / In Portugal, cork oak cultivation goes hand in hand with agroforestry: Iberian sheep, cattle, and black pigs feed, in part, on the acorns that fall to the ground as well as the grass, herbs, mushrooms, and even grains that can be found beneath the trees.
© Daniel Michalik

3

/
Botanical Properties

1 / A forest in the south of France
© Mediterranean Institute of Cork

2 / The bark of the cork oak, seen here under a microscope, is formed of small hexagonal or pentagonal cavities that enclose an air-like gas. These cells measure between 10 and 50 millionths of a millimeter and are organized perpendicularly to the trunk of the oak. Cork is essentially composed of this gas and *suberin* (a kind of natural wax contained in the walls of each cell, which is what makes cork waterproof). The gas enclosed in these cells makes them light and insulating. A plank of cork taken from the trunk is composed of 60% gas. The cells are essentially like little cushions, and this is what gives cork its elasticity; it can also be compressed by half without losing any flexibility.
© Amorim

3 / It's easy to see the dividing line between the original bark of the tree and the bark that regrows after the first debarking.

3

4

5

6

/
Botanical Properties

4 / The bark of the tree is the material that is called *cork*. Thick and filled with crevasses, this bark can grow up to 10 inches (25 cm) thick. In addition to reducing water loss linked to drought, the bark also provides a natural protection against the frequent fires that occur in cork-growing regions. The thick bark ensures that fire can't reach the *sapwood*, the living part of the tree beneath the bark, so new branches are able to regrow quickly whereas other species exposed to fire die. Because of this, the tree is used today to help fight fires.

5 / A forest in Portugal

6 / The leaves of the cork tree are small, tough, and evergreen. The tree blooms in the spring with yellow flowers that then become oval acorns. The heartwood is dense and very hard, a marvelous heating wood that is also used in woodworking and small carpentry projects.

CULTIVATING CORK

The bark of the tree is removed to produce cork. This manual operation is called *debarking* and must be done carefully by someone with the right knowledge so the tree is not damaged. If the process is done carefully and successfully, the bark will regrow. The bark is removed once every nine years at the earliest, between May and August, during a phase when the tree is most actively growing. The slower the growth, the denser the cork.

The first debarking, called the *unmasking*, is done when the tree is about twenty-five years old and results in what is called *virgin* cork, which is highly irregular and very hard. This first material cannot be used to make corks but is instead crushed and made into other products like insulation or floor coverings.

At the second debarking, the cork removed is still irregular, less hard, and of average quality, but it is still not ready to be made into bottle corks. When the tree reaches its third harvest, a minimum of eighteen years after its original unmasking, the cork is finally ready to be made into bottle corks; this is called *reproduction cork* and it is smooth and regular.

Every nine years for the next one hundred and fifty years, the same process is repeated, resulting in approximately sixteen harvests over a tree's lifespan.

1

2

/
Cork Trunks

1 / 2 / The initial bark of the cork oak is filled with crevasses; when removed, it is called *virgin cork* **(2)**. Once this first layer of bark is removed, a second layer grows naturally to cover the exposed trunk. The second layer of bark is smoother and more regular, and it is called *secondary cork* **(1)**.
© Mediterranean Institute of Cork

3 / A tree after cork debarking. The bark is removed all the way up to the base of the first branches.
© Mediterranean Institute of Cork

12
3

1

2

3

4

/
Debarking

1/2/3/4/5/6/7/
The expert who removes the bark opens the tree vertically with a special axe in order to mark out the future planks. He must twist the axe to unstick the bark from the trunk. He places the axe handle between the belly of the plank and the inner bark (which must be preserved so the tree will continue to grow) and disengages the plank.

He marks the top edge of the desired strip with a horizontal cut and removes the plank. The height of the plank determines its worth.

5

6

7

1

2

3

4

/
Initial Processing

1/2/3/4/5/
Following the debarking, the planks are stored in the open air for a minimum of six months until they stabilize and mature. Sorted according to their visual quality and thickness, they are then boiled for an hour to help the cork expand and acquire its mechanical properties: the goal of this stage is to disinfect and clean the material, to extract its water-soluble substances, to increase the thickness of the planks, and to reduce their density, all while making them more supple and elastic. The gas contained in the cells dilates and increases the volume of the cork by 20 % in addition to making it more even. The cork is then stabilized and flattened for two to three weeks before it is sorted and placed into the production chain.

5

THE HISTORY OF CORK

In ancient Egypt, cork was used to make commonly used objects like dishware or fishing bobbers, and cork stoppers were used in ancient Greece to close amphoras and bottles. During the sixteenth century, the industry saw its first upsurge in Europe when Dom Pierre Perignon began to use cork to bottle his champagne. Beginning in the nineteenth century in Europe and the United States, new products like composite cork and white cork were developed for use in flooring production. In the following century, cork began to be used more and more in manufacturing. Its insulating properties made it of particular interest for building cold storage. Slowly, however, plastic —cheaper, but less durable—began to replace cork in numerous applications and as a result, cork cultivation has been on the decline in France since the 1950s. Cork production continues in countries with cheaper labor forces.

1

2

/
A Bottle Cork Unlike Any Others

1 / A bottle cork manufacturing workshop in the 1940s
© D.R.

3 / 5/ Bottle corks are produced directly from the boiled, stabilized planks. The cork is cut into strips just barely wider than the length of the bottle corks needed.
© Amorim

4 / A kind of cookie cutter is used to cut the corks from the cork wood strip, just beneath the bark's gray surface. Today this work is mechanized and completed using a puncher that creates the cylindrical shape of the bottle cork. Nothing is lost in this process because any waste is ground up to be used in composite cork or as an additive.
© Amorim

2 / 6 / Bottle corks are sorted according to their appearance and porousness. Today, automatic sorting has mostly replaced visual sorting. Finally, the bottle corks are washed, disinfected, dried, and eventually sealed (the cork pores are filled with a mixture of glue and cork powder to make them more waterproof).
© Amorim (2)
© IML (6)

3

4

5

6

TECHNIQUES

Cork has recently made a big comeback, especially in the industries of building, fashion, and furniture design. Novel applications have appeared, and the material is worked and transformed in increasingly inventive ways into a variety of objects and materials.

Cork is a material of choice in numerous applications because of its exceptional properties: it is low-density, antistatic, and acoustic; it is a heat insulator and shock absorber; and it is fire resistant and waterproof. It is also soft to the touch, rot resistant, and infinitely recyclable.

Cork can be transformed in many ways. After it is boiled and stabilized, natural cork undergoes no further processing and is primarily used to make bottle corks. The byproducts of this process are ground and sifted and used in forms for building works or to make white composite cork or expanded black cork; powdered cork can be combined with additives to form various composite materials like plastic agglomerated cork. It is even possible to transform cork into paper, fabric, or something that looks and feels like leather.

1

/
An Extraordinary Material

Cork remains stable over time because it is resistant to condensation and mildew, does not decay, cannot be destroyed by vermin or atmospheric elements, and is tolerant of high temperatures and resistant to wear.

Cork's numerous finishing possibilities enable designers to create products to suit their demanding portfolios. After designers discuss their ideas with the manufacturer, prototypes are made from their project sketches until the manufacturer produces a shape that embodies the optimal balance between user experience and the constraints of production and sales. A series run follows, thus initiating the industrial production of the product in its final form.

1 / Agglolux,
© Jean-Baptiste Coffin

2 / Stow It project, design by James Irvine
© Amorim

2

/
Initial Processing

1 / Grinding waste cork
© Amorim

2 / Manufacture of expanded black cork, a 100% natural and sustainable project. Black cork is an exceptional insulator that can handle wide temperature fluctuations. Once the cork is ground and dried, it is heated at a high temperature in an autoclave that injects water vapor at over 570° Fahrenheit (300 °C). At this point, the air expands, which makes the cork lighter and even more insulating, and it takes on a dark hue. Because the suberin within the material acts as a glue, there are no additives. The agglomerated blocks are then cut into planks of different thicknesses and packaged.
© Amorim, photography by Daniel Michalik

3 / From a carbon standpoint, this is "green" production because much of the heat needed for the process is produced using woodchips and residues from the production itself—biomass is burned in order to keep the factory working.
© Amorim, photography by Daniel Michalik

4 / Storage of approximately 3' × 3' × 1.5' (1 m × 1 m × 50 cm) blocks of agglomerated expanded cork. Boards will then be cut from these blocks.
© Amorim

5 / Woodchips used in manufacturing expanded (black) cork and composite (white) cork
© Agglolux, photography by Jean-Baptiste Coffin

1

2

3

/
Machining

1 / 4 / 5 / Composite cork can be worked using the same tools as wood: it is sawed **(4)**, drilled **(5)**, sculpted using a lathe **(1)**, and milled either by hand or using a five-axe robot to carve complicated shapes into the material.
© Amorim

2 / 3 / White composite cork is generally more fine-grained. The ground cork is agglomerated into blocks (here, cylindrical), which will be cut and reworked using different glues depending on their future use (alimentary, acoustic, for outside use). When manufacturing joints, protective cushions, and orthopedic soles, rubber can be added to the initial mix to improve its shock-absorbing properties. Cork can also be directly molded into a final desired shape. Agglolux makes products out of cork in Landes, France.
© Agglolux, photography by Jean-Baptiste Coffin

4

5

3

4

5

/
Fabric and Paper

1 / Fabric and paper are obtained by gluing or hot-pressing a thin sheet of cork or cork particles onto a textile background or paper. The material is then rolled onto a spool.
© Amorim

2 / Rolls of dyed cork fabrics by 4Spaces
© 4Spaces, www.4spaces.ch

3 /4 / 5 / Depending on the textile background used, the fabric has different strengths. It can be used for furniture, like this foldable chair from Pearl Cork, or to cover or decorate an object—even to give it the feel of leather, like this helmet and ball from Sedacor.
© Sedacor, an imprint of JPS Cork Group (3 and 4)
© Pearl Cork, www.pearl-corkdesign.com (5)

6 / Wallpaper with various prints and colors made a splendid geometric décor that set the stage for the exposition at the International Fashion Showcase at Somerset House for London Fashion Week 2016. Different papers from Wicanders' Deckwall collection were used.
© Bloom Portugal Fashion, exhibition curator Miguel Flor, scenography by Miguel Bento Wicanders, www.wicanders.com

6

7

8

/
Fabric and Paper

7 / 8 / The "natural" collection of cork fabrics from JPS Cork Group includes white and black cork. The hues here are from the material itself, but other collections are dyed or sculpted.
A wide variety of decorative styles is obtained by using cork that is first cut into strips, squares, or other shapes and then recomposed on a fabric background.
Samples from other collections: dynamic cork, cork wall tiles, cork wall rolls, cork wallpaper
© Sedacor

9 / The Swiss-based textile designer 4Spaces offers a collection of furniture fabrics and draperies in cork. The fabric collections include multicolored prints, which for some models are combined with a texture or laser cut.
© 4Spaces

10

11

12

/
Fabric and Paper

10 / 11 / 12 / 13 /
A cascade of magnificent drapes intended for interior and window design.
This collection of cork lace fabrics combines different patterns and sizes of cutouts. Some of the cork fabric models offered by the Zurich-based company 4Spaces are laser cut. These delicate lacy fabrics have geometric or floral motifs and, depending on the model, show the fabric background onto which the cork is glued. In this way, the oak bark becomes a kind of brocade.
© 4Spaces

1

2

3

4

/
Composite Materials

Waste cork that is finely ground into cork powders can be used as an additive to concrete in industrial composites, for the manufacture of linoleum, or even when making the substrate for soccer fields. The diversity of additives, binders, and other mixes —resins, plastics, rubbers—makes it possible to create a wide variety of products in all shapes and sizes.

1 / Surfboard
© Amorim, Mercedes-Benz, and Garrett McNamara

2 / CorkFill cork filament for 3D printing by colorFabb and objects printed from cork; "Parametric Twist" vase, design by gCreate; "Tower," design by seechless.
creative commons, www.colorfabb.com www.gcreate.com www.thingiverse.com/thing:99028 creative commons, www.colorfabb.com

3 / This shoe is flexible and antibacterial and is meant to be worn by people who usually walk barefoot. The combination of silicone and cork powder is the key to this composite product that won't irritate feet.
© Barefooters

4 / Floor panel for a train car
© Amorim

5 / **6** / Manufacturing light concretes using cork powder as an additive
© João Luís Carrilho da Graca for Metamorphosis, Amorim; (experimentadesign), www.experimentadesign.pt

5

6

1

2

3

4

5

/
Insulators
Floor Coverings

1 / The benefit of using cork shavings for heat insulation in attics or wood frame walls is that they don't compact over time. Cork is the best thermal and acoustic insulator. Its stability over time is basically limitless. Like other insulators, it conducts heat, but as shown in a study by CRESTEB lab for the Bâtiliège company, its properties are altered much more slowly than other insulators when confronted with extreme cold or heat. It is waterproof but not vapor proof, so it allows walls to "perspire." The excellent density of cork insulator panels also resists compression.
© Amorim

2 / Roll of linoleum "Divina Corkoleum" by Sedacor
© Sedacor, www.divinacorkoleum.com

3 / Wicanders laminate parquet is composed of an underlayer of either composite cork or wood and a cork or wood finish.
© Wicanders

4 / 5 / Cork is used to insulate under roofing, between brick walls, and beneath flooring, or it is placed inside or outside a wall. For the latter application, it is either coated or left as is like this facade cladding from Amorim. Placed as a thin underlayment beneath a parquet floor, it reduces the sound of footfalls as well as other noises.
© Amorim

6 / 7 / The absorbent qualities of cork make it an interesting choice for correcting and insulating a room's acoustics. Placed against a wall, the cork panel can be coated or left visible when the surface is sculpted, shown here in Amorim's MD Facade Wave collection.
© Amorim

6

7

7

8

9

/
Floor Designs

7 / Custom poured cork floor composed of cork powder and natural resin. This technique works for both interiors and exteriors. The floor is both flexible and warm.
© L'écorce du Chêne, Christophe Crublet, www.revetement-liege.com

8 / The floor of the Sagrada Familia in Barcelona was covered with approximately 21,500 square feet (2000 m^2) of cork comfort parquet from Wicanders before the Pope's visit to consecrate the basilica in 2010. The cork was able to be installed much more quickly than a marble floor, and at a lower cost. In addition, it is highly durable and looks very much like stone.
© Dr. Jordi Bonet, chief architect for the Sagrada Familia, who is continuing the work that Antoni Gaudi began in 1882

9 / Mosaic floors for bathrooms and exteriors make use of industry scrap materials and are produced entirely by hand using a social integration company.
© Friedbert Bleile Handel und Dienstleistungen, www.korkmosaik.com

10 / Wicanders floor in laminate parquet for the Belle Etoile project—a children's home in Pierrefeu-du-Var
© Sandrine Cœur-Bizot

11 / A cork floor is apt to be highly appreciated in companies that have a lot of employee foot traffic. Its softness absorbs shock and reduces muscle fatigue. The variety of designs offered by the Cork My Way and Divina Cork Floors collections from the JPS Cork Group illustrate numerous possibilities for interior design.
© Sedacor

10

11

1

2

3

/ Objects of Daily Life

1 / 2 / Composite white cork with food-safe glue can create water-proof and food-safe cooking utensils. This little boat floats because of its low density.
© Daniel Michalik

3 / Magic box
© Black Cork, designed by Rui Pereira

4 / Cork's insulating properties make it useful for conserving both heat and cold, like this set of teapots called "The Whistler."
© Amorim collection, designed by Raquel Castro

5 / 6 / Cork is soft to the touch and its flexibility makes it a material of choice for furniture design, as seen in these stools. The expanded black cork stool from Blackcork **(6)** uses the same principle as the corkscrew— the feet can be screwed into the seat. The second stool, in white composite cork from Amorim, is sufficiently thick so that the feet can be nested into the seat, as shown here **(5)**.
Senta stool, Materia collection **(5)**
© Amorim
Fernando Brizio Rolha coffee table **(6)**.
© Blackcork, designed by Goncalo Campos

4

5

6

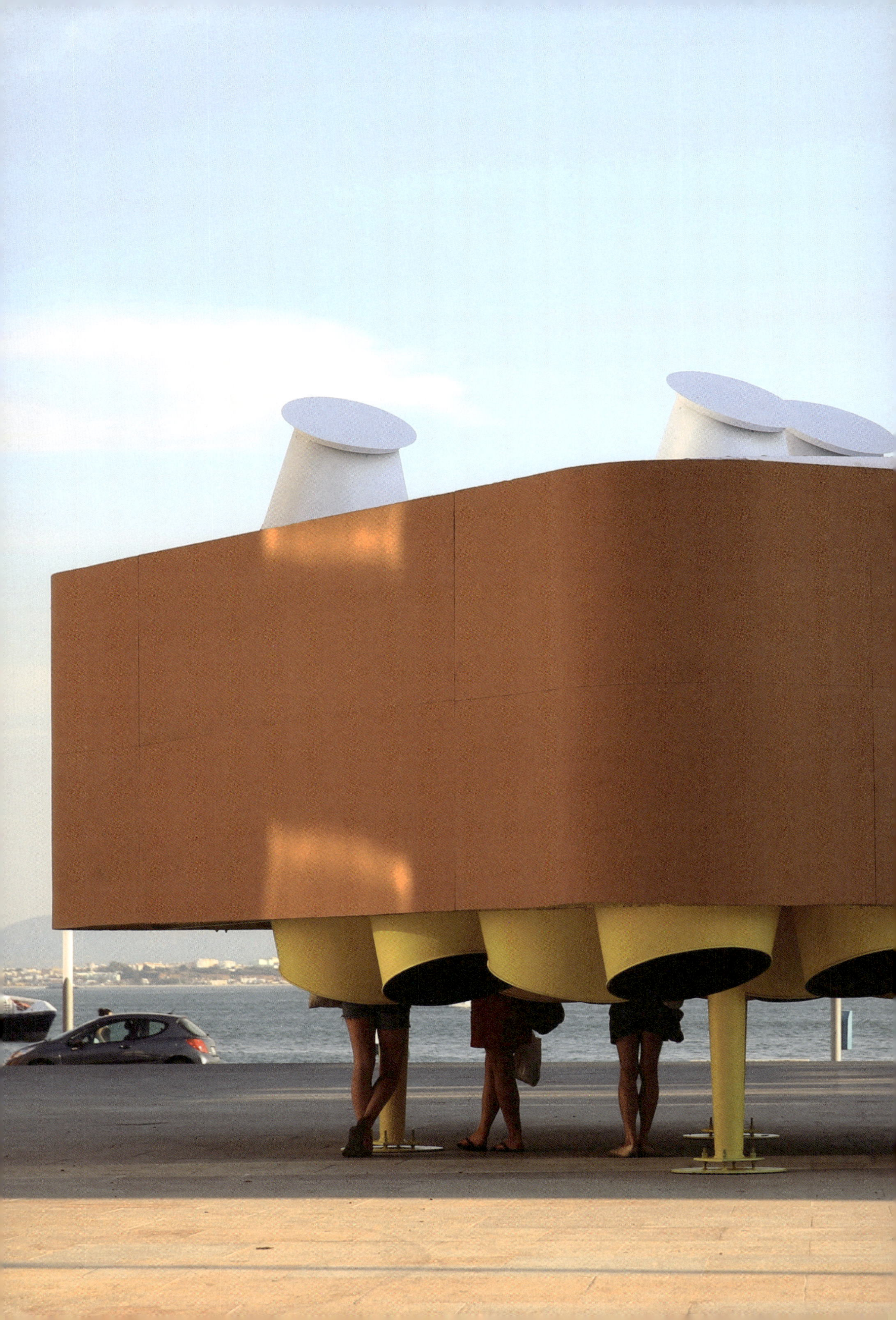

CORK in Architecture

/ “Open Cinema,” by Colin Fournier and Marysia Lewandowska. Cork is an exceptional material and can be used in the most extraordinary architectural creations. CC N-C-S-A v.3.0, Colin Fournier and Marysia Lewandowska

/ Cork floor covering, Cork My Way collection © JPS Cork Group

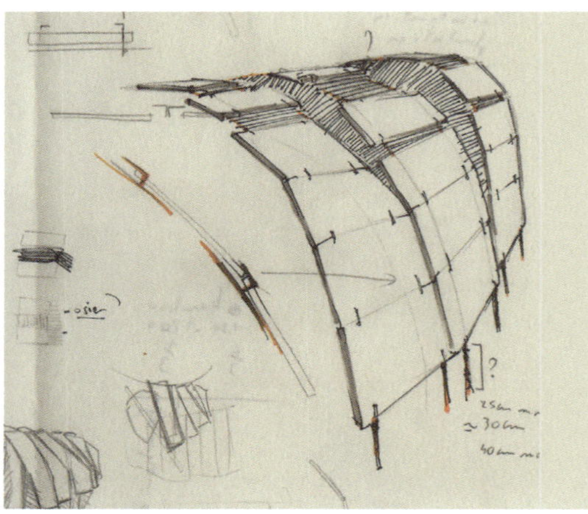

This structure was produced using rebar covered with expanded cork chosen for its acoustic qualities. Blue wicker has been braided like a kind of stitching between the curved panels of this organic shape.
© Concepteurs

Acoustic Landscapes / Olivier Brichet, Alexandra Epée, Flora Rich

/ "Gram(in)ophone" installation
/ International Garden Festival, Chaumont-sur-Loire, 5th edition "Jardins des sensations—synesthésies contemporaines," 2013
/ domaine-chaumont.fr
/ Olivier Brichet (scenography, builder, and sound designer), olivierbrichet.com
Alexandra Epée (designer, scenographer), lule-studio.com
Flora Rich (landscape architect), florarich.wordpress.com

Each year since 1992, the Chaumont-sur-Loire Manor has been used as a "laboratory of contemporary landscape architecture." In 2013, Brichet, Epée, and Rich were selected to create an ephemeral themed garden; the scenographer, sound designer, and landscape architect decided to create an immersive experience as a "synesthesia." The result was an acoustic walk through a garden of grasses. The vegetation was dense at first and then it opened up to reveal intriguing architecture. In the dark enclosure with its muted sounds, vibrating panels provided visitors with all kinds of amplified sounds coming from the surrounding microcosm.

The designers chose their materials carefully: recycled telephone poles for the stilts, cladding made from locally sourced cedar, cork for insulating the walls and ceilings.
On the tilted side panels, the cork is apparent as cladding, and the interior of these panels is lime plaster. The floors are insulated with recycled cotton fibers.
© Henrietta Williams

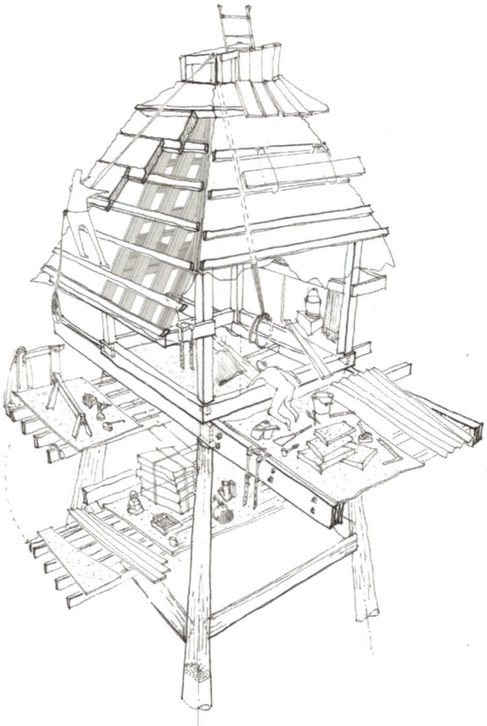

Life in the Woods / Nozomi Nakabayashi

/ "Hut on Stilts"
/ England, 2015
/ www.nozomi-nakabayashi.com

Somewhere in the Dorset woods, a writer dreamed of an inspired retreat setting. The suspended cabin, designed and built for the writer by Nozomi Nakabayashi along with a small team of builders and basic tools, is a testament to the fine skills of this young graduate of the Architectural Association school. The small hut was conceived of as a house containing only the barest necessities. It includes clever built-in furniture and storage and has a wood stove. The living space, organized into two half-levels, opens out onto a small terrace. The windows frame a landscape that invites meditation.

Using cork cladding provided by Amorim, the engineers and architects created parametric 3D models for an experimental cork structure to minimize losses and to be able to use the material as a self-bearing structure. Wooden supports stabilize the entire construction.
© Maleczek, Collectif Archiwaste / koge–University of Innsbruck | Department of Design | Structure and Design

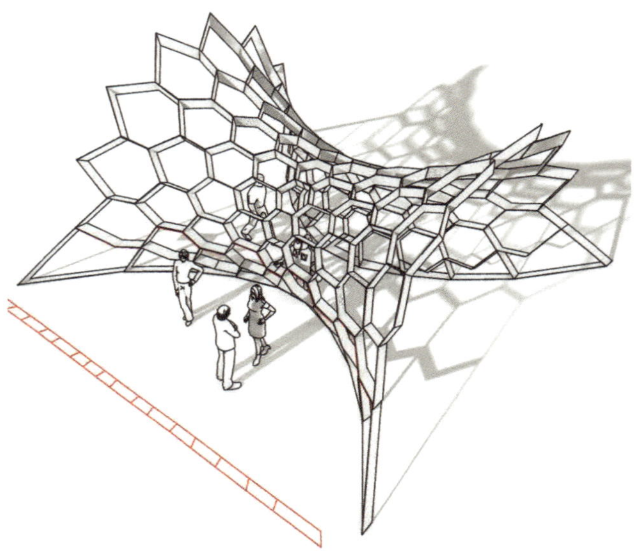

Rupert Maleczek, Archiwaste
Christian Scheiber, koge

Experimental Cork Structure

/ Negamat Workshop for Vivexpo
/ Vivès, France, 2014
/ www.archiwaste.org
/ www.archispass.org
/ www.koge.at
/ www.institutduliege.com

In the Eastern Pyrenees, the use and production of cork has lost its traditional dynamism. This means there are local challenges to maintain the cork oak forests and the activities traditionally linked to this crop. Local institutions decided to organize an experimental field so that this traditional material could be perceived anew as innovative. They succeeded through a cultural exchange with some Austrian researchers: a cork net was built by designers, students, and volunteers on the main village square in Vivès during the International Cork Festival.

1

The pavilion is entirely covered with cork on both the inside and the outside.
© Gabriela Gomes (1, 2, 3, and 4)
© João Morgado (5)

2

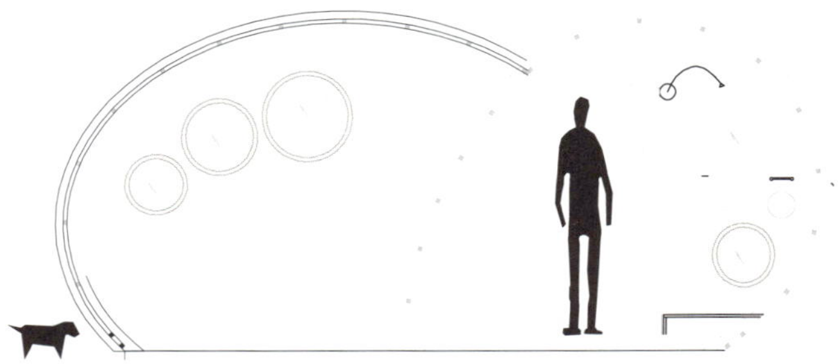
3

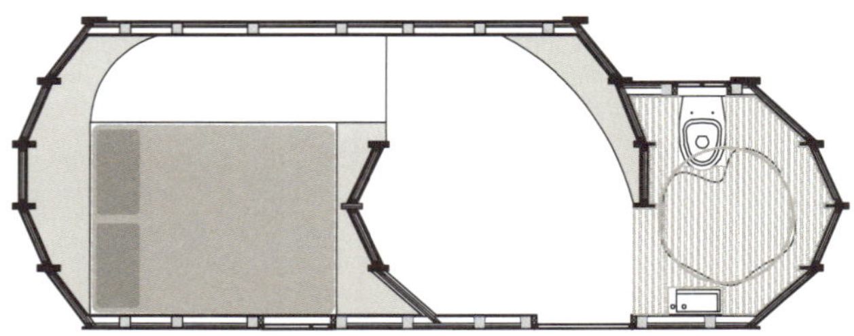
4

A Shelter in the City / Gabriela Gomes, Sergio Bessa

/ "Shelter" by GG, livable module
/ Guimarães, Portugal, 2012
/ www.shelterbygg.com

Public spaces are increasingly used by cultural events not bound by traditional labels, a sign of changes in cultural mores. This nomadic shelter is a design project, a piece of architecture, a shelter and, as a bearer of messages, an art installation. From an artistic perspective, it's a kind of multicolored slice of the city and also an environmental manifesto because of its choice of materials (like cork) and renewable energies. Finally, this mobile work uses cork to pose the question of whether it's possible to provide private space within a public space.

1

2

3

Open Cinema

Colin Fournier, Marysia Lewandowska

/ Open Cinema, Installation / Film
/ Guimarães, European Capital of Culture 2012 Lisbon, Architecture Triannual 2013
/ Acknowledgements: DOF Cork

Public spaces have long been reduced to their external function as urban facade, but today, architects, urban landscape artists, sociologists, and artists work together to create living maps of something experienced from within. Setting an anchor in a public setting enables passersby to stop and enjoy a welcoming space, a place in which they can be distracted or take refuge from the bustle of the city. This open-air cinema is not only an artistic performance space, but it also invites reflection on the memory of place and the social moment that occurs within.

4

5

6

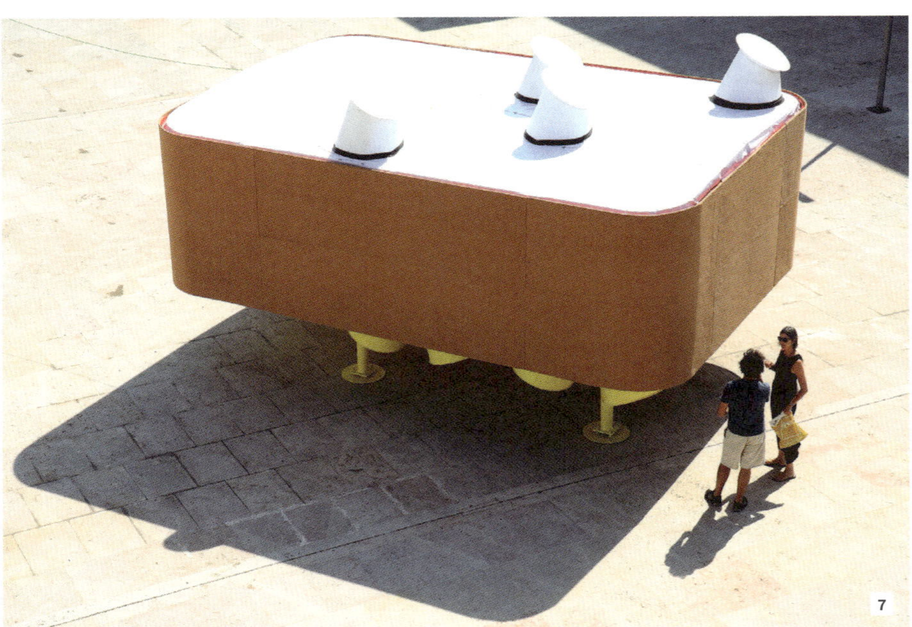

7

8

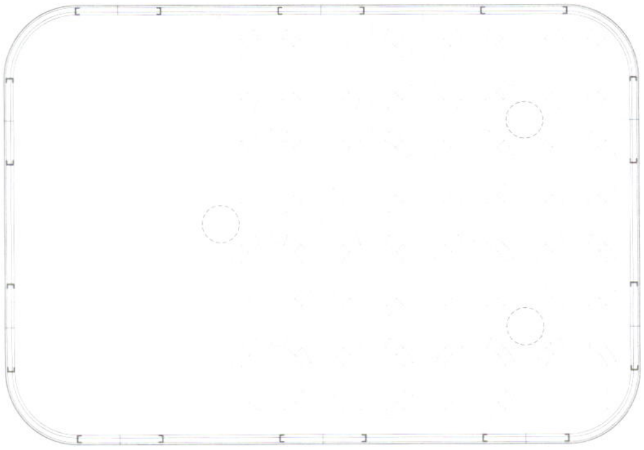

9

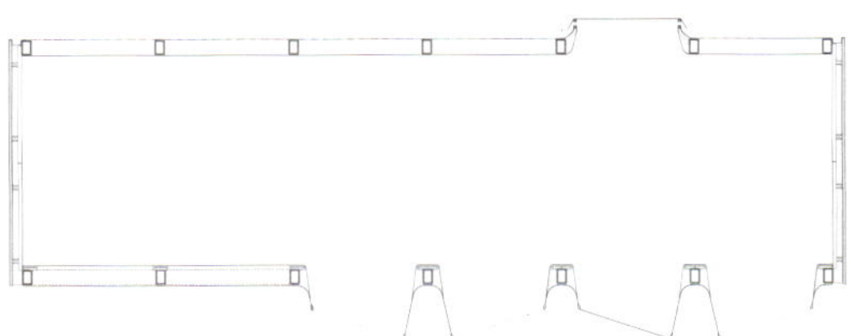

10

1/2/3/4/8/
The open-air cinema is entirely collapsible and covered in cork both inside and out. Viewers enter a kind of cocoon that mutes external noises and is warm in the winter and cool in the summer—all of which favors an immersive experience. Projections were from cinematographic archives on the resistance to the Portuguese dictatorship under Salazar.
CC N-C-S-A v.3.0, Colin Fournier and Marysia Lewandowska

5/6/7/
Setup stages
CC N-C-S-A v.3.0, Colin Fournier and Marysia Lewandowska

9/10/ Diagram and crosscut view
© Colin Fournier

1

3

2

1/2/3/4
When visitors analyze the in-ground portion of the site, they can isolate a series of layers and see that their overlap is what makes the complex constructive framework of the project work.

Visitors note the circular roof that integrates a pool that reflects the sky; beneath that is a textured landscape in black cork selected for its tactile and olfactory qualities.

© Herzog and de Meuron and Ai Weiwei (1 and 3)
© Antonio Moll (2)
© Iwan Baan (4)

Back to Origins / Herzog, de Meuron, Ai Weiwei (1 and 3)

/ Pavilion of the Serpentine Gallery
/ Kensington Gardens, London, United Kingdom, 2012
/ www.herzogdemeuron.com
/ www.aiweiwei.com
/ www.serpentinegalleries.org

Each year, between 2001 and 2011, talented architects were invited to construct a summer pavilion for the Serpentine Gallery in Kensington Gardens. After so many illustrious predecessors, the designers of the 2012 pavilion were faced with the inevitable question of form and continuity. They decided to sidestep this formal question by digging a hole, a kind of well, until they reached the water table. Inspired by the tangle of vestiges—subterranean networks, foundations of previous pavilions, and other elements—the project was conceived as an archeological dig enveloping and covering the entwined memory of the pavilion's previous incarnations.

4

Because patrons of this exhibition —visited by 60 million people— were incredibly curious about the cork used, the organizers were forced to change thousands of panels! Carlos Couto tells us: "I remember that one day we met an old Chinese lady who was carrying an entire cork panel away with her. It was awkward at first, but after some thought, I let her leave with her trophy ... she deserved it. She took my pavilion ... that's quite a souvenir!"
© CC, Atelier de Arquitectura

A Place for the World / Carlos Couto, Atelier de Arquitectura

/ Portugal Pavilion
/ World's Fair, Shanghai, 2010
/ www.ccofficemacau.com

The theme of the pavilion was "Portugal, a place for the world," and so it needed to create a meeting space, a place to stroll at the crossroads of culture. Portugal had a prefab pavilion; to cover its rigid base, the architect and his team designed a volumetric facade inspired by lines from the sailing maps so cherished by the Portuguese people. This "relief map" was covered with a completely new material for many Chinese tourists: Portuguese cork. This renewable material sent a powerful message and turned out to be an attraction for the public; large crowds came to feel and touch it and even to cut off small bits of the facade to take away as souvenirs!

葡萄牙
PORTUGAL

The cork facade strikes a contrast to the lime plaster walls of this contemporarily designed building that integrates easily into its natural and cultural surroundings. The cork cladding glued to the masonry keeps the building cool in the summer.

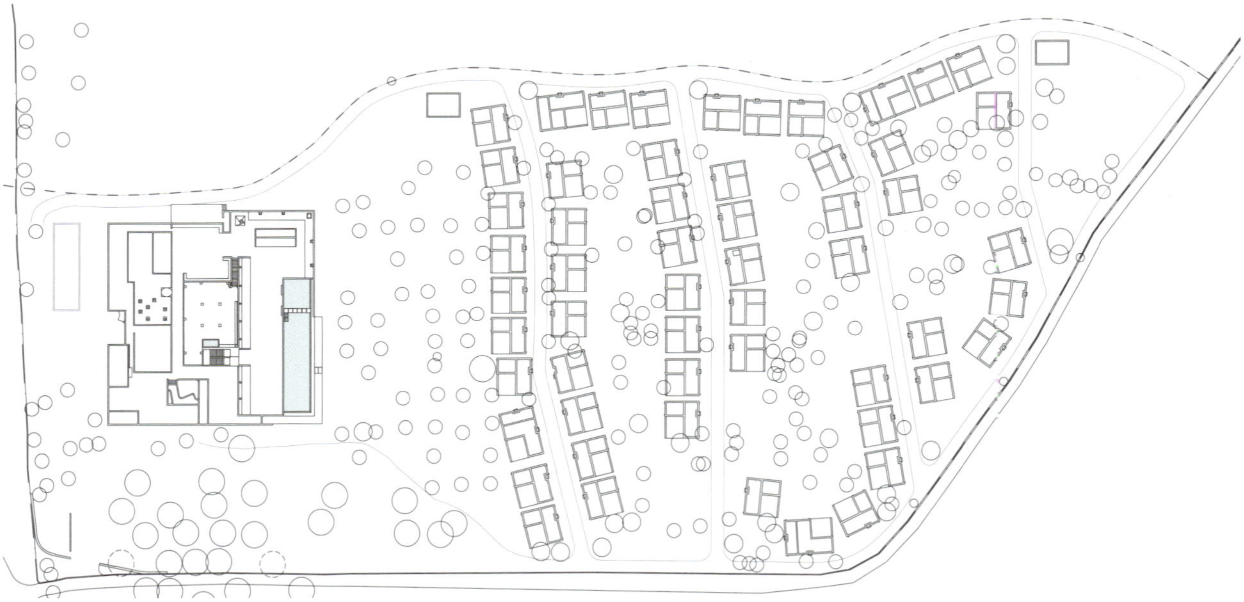

Hotel of Dreams / José Carlos Cruz Arquitecto

/ Ecork Hotel
/ Evora, Portugal, 2013
/ www.josecarloscruz.com

Imagine the Portuguese countryside: olive trees, cork oaks, a typical road of the region of Alentejo, fragrance, and the contrast of shadow and light ... you've just arrived in a dreamscape inspired by the vernacular Arab architecture of the Middle Ages. The main building of the hotel complex is like a stronghold: a simple, large space organized around a central courtyard and surrounded by smaller dwellings that are actually independent suites. From the roof, you can see the subtle play of the shapes in their configuration beneath the light; the pool and the magnificent countryside will have you forgetting everything, even the reason for your travels ...

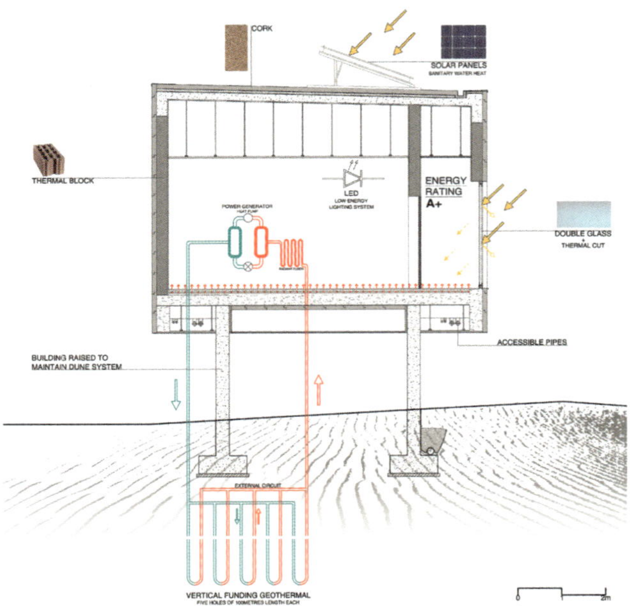

1 / 2 /
The pure-lined shapes were covered in 2.75-inch-thick (7-cm-thick) cork facade cladding panels and glued onto the heavy structure. This design keeps the buildings cool in summer.
© José Campos

3 / Like a wading bird, the construction picks its way across the dunes. Once demolished, the site will be easily restored to its original state. Minimizing the impact of construction on the environment is the future of architecture.
© José Campos

Good Vibrations ⁄ Branco Cavaleiro Architectes

/ Surf High Performance Center
/ Vianan do Castelo, Portugal, 2012
/ www.cavaleiroarchitects.com

For this project, the architects were challenged to build a structure that would house thirty high-level surfers and their specialized personnel in a stunning, but fragile site. The center's buildings had to be as high performing as their guests. When building inside a natural refuge, it's vital that the architecture that is conceived is as ephemeral as its location and that the architect's plan for its eventual dismantling will have as little impact on the site as possible. This U-shaped construction is settled delicately between the trees, steps over the dunes on stilts, and blends into the landscape when the beige color of the cork cladding is lit up by the sun.

1 / 2 / 3 / 4 / 6 / The "double-skin" cork covering for the building was made of 2-inch-thick (5-cm-thick) facade cork cladding by Amorim. It was fastened and glued onto a metallic cladding held up by a steel structure.

5 / A more recent project by the office, which looks similar

© N'engenharia

Covered in Cork / Manuel Gonçalves, Jaime Pinto Gomes, N'engenharia

/ Renovation of the AEP Exhibition Park
/ Associação Empresarial de Paços de Ferreira, Portugal, 2014
/ Architecture Manuel Gonçalves and Jaime Pinto Gomes
/ Engineering: N'engenharia, www.nengenharia.com

The Exhibition Parks designed during the second half of the twentieth century are generally immense complexes without any architectural unity; they are "drunken boats" with vast stretches of parking lots.

Such parks are difficult to renovate because of their size and how spread out they are. Contemporary designers must work not only to correct the spatial perceptions of visitors who can feel lost in such large spaces, but also to improve energy performance and regulate heat. Re-creating a unified whole through a cork skin that covers both the older and the newer constructions is a promising response to the problem of "energy loss catastrophe."

6

The building carefully alternates facade cladding panels from Amorim—which also cover the roof—with perforated metal shutters.
© Ivo Canelas

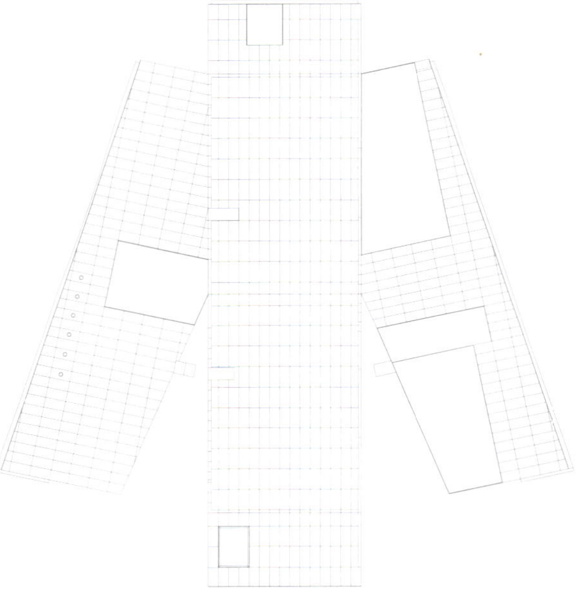

Simple and Economical

/ Arquitectos Anónimos and Paulo Teodósio

/ Cork House, private home
/ Esposende, Portugal, 2007
/ www.arquitectosanonimos.com

Designing a private home is a complex job for architects. On the one hand, budgets are limited and the size of the project doesn't allow for economies of scale, and on the other hand, the house as a symbol of success can invite irrational projections. The work entails supporting the house's future inhabitants in stripping away everything that is unnecessary, leaving only the right shape, the one that, despite appearances, will perfectly meet the family's needs. This is essential when the budget is limited. The architects used this very simple cork structure, which suited the project's reduced budget, to parallel the simplicity of means with a purity of shape.

1

2

3

1/2/3/4/
This building is covered with cork facade panels that are used both as cladding and as insulation that is glued from the outside directly onto the hollow brick used as load-bearing walls. The project also integrates the topographical curves of the land-scape—so much so that from the road, the building looks like an earthen terrace or a fifth wall. And looking from the garden, on the river's edge, there is a play of forms that is like cork furniture on an architectural scale.
© Contaminar Arquitectos (1, 2, 3 and 4)
© Fernando Guerra (5)

Leaning on the Hill / Contaminar Arquitectos

/ The Hill Cork House, private home
/ Leiria, Portugal, 2007
/ www.contaminar.pt

Topography is often neglected in architectural projects. Everyone would like to build the highest tower to overlook the neighbor's roof... it's rare to find those who can work at the waterline. Finding a humble position beneath something, in the way a river does, is a different kind of challenge. This private home is an excellent example of integration: its rear wall and its rooftop garden that includes pathways seem to melt into the vineyards. The living space opens out onto a lower garden. The house is divided into two "blocks," separated by an indoor pool, a design that keeps the bedrooms separate from the living spaces.

4

5

1

2

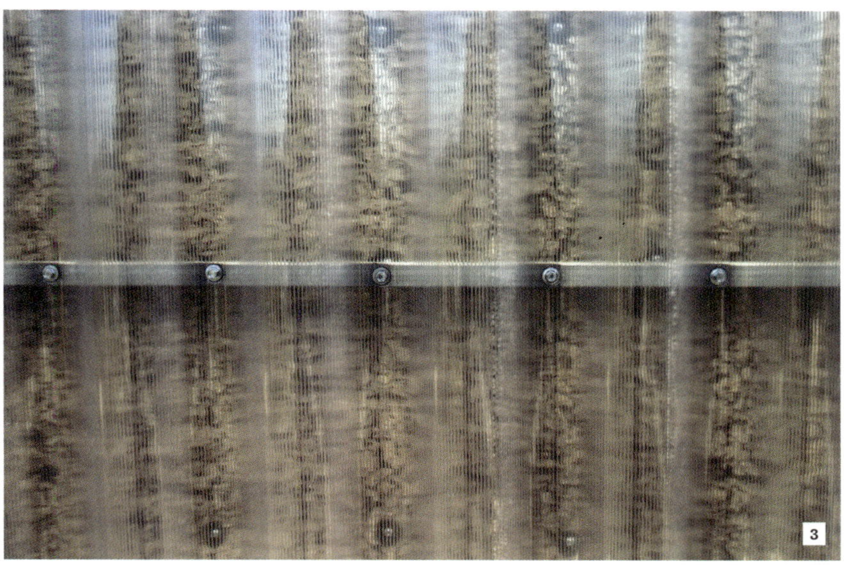

3

1/2/3/
The main structure is made out of eucalyptus and the walls are covered in two layers: 2-inch-thich (5-cm-thick) cork and a clear, but ridged, polycarbonate. The gaps between the frame are filled with a mixture of earth and straw. Finally, a lattice of planks works as a support for a finish of plaster and whitewash.

The entire design is extremely energy efficient and the selection of materials ensures the air quality inside the home.
© Sandra Pereira, photography

Système D / Plano B Arquitectura

/ Wattle and Daub House, private home
/ Leiria, Portugal, 2007
/ www.planob.com

In a green-belt-classified zone, it's forbidden to erect new construction. Nevertheless, when a house falls to ruin, it is possible to rebuild without changing the location of the reconstruction. For this project, within an imposed footprint, the architects did not have much latitude. They decided to use traditional materials in a modern way. When finished, the building looked like a modern agricultural storage shed. The combination of current building techniques and natural raw materials, like the stone of the foundations, wood, cork, and earth, is an economical and coherent strategy from a contemporary art of living and building perspective.

This small construction site is also very educational—both from an architectural perspective and in terms of the environmentally responsible building techniques that the architects from Plano B hold dear.

4 / Upon the former foundations, and using stones recycled from the preexisting ruin, they created a masonry platform and the support necessary to access the construction.

5 / From this point, they began to erect the building and had enough space to organize the work site. Wooden frame walls were built and crossed with diagonal beams to ensure the stability of the whole. Metal pre-frames were positioned to support future woodwork.

6 / 7 / The building was weatherproofed: the frame was covered with cork and filled with a mixture of earth and straw, a kind of contemporary cob. The woodwork was placed.

8 / A ridged polycarbonate cladding was fastened onto the battens before the cork, leaving the latter visible. Wooden sliding shutters ensure protection from the sun and any break-ins. The roof doubles as a terrace and all its parts—base, cover, and rivets—were done in sheet metal to make it waterproof.

9 / © Plano B Arquitectura

8

A. Cob
B. Cork panels
C, E, G. Metal
D. Polycarbonate
H. Roof / terrace

1. Platform
2. Living room
3. Kitchen
4. Storage
5. Utility room

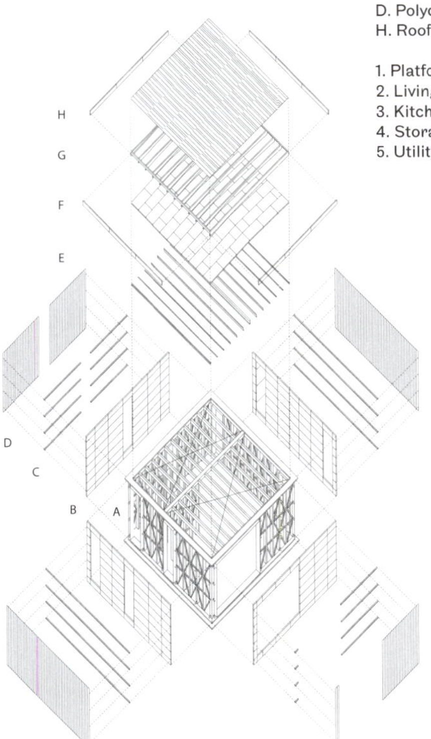

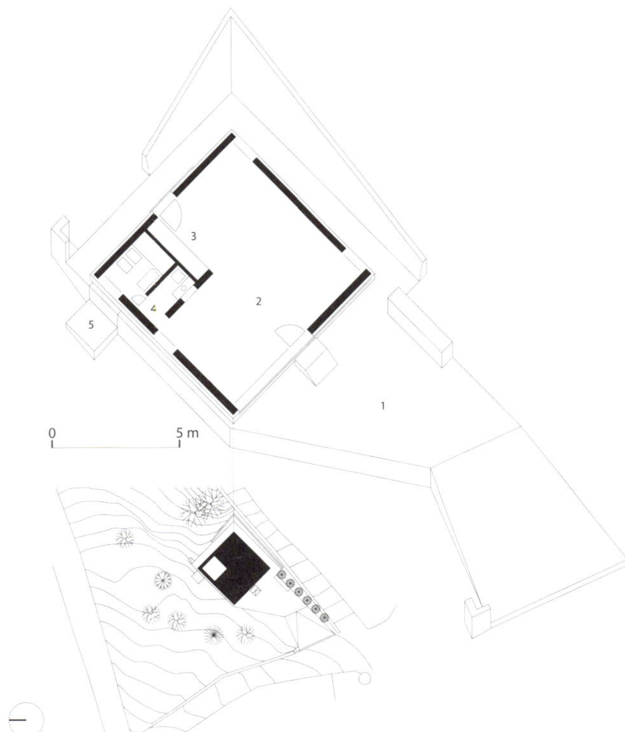

9

The prefab wood system, covered in cork, is entirely transportable and packable into a single piece. The cut and stacked plywood panels make up a framing system optimized for digital cutting. The modules can be multiplied using the geometry of a tetrahedron. These kinds of expandable, light constructions are called *modular architecture.*
© Tetra Shed

Innovation Imperative

Housing Units

/ Tetra Shed, prefab module
/ London, England, 2011-2014
/ www.tetra-shed.co.uk
/ www.tetrahotel.com

There is great demand for small, contemporary-design home extensions. This modern "garden shed" can work as an office or guest room. It's a housing concept taken directly from modern architecture. It garners much attention in its updated contemporary form. The uniqueness of the patented and trademarked Tetra Shed system is the combinatorial possibilities that are enabled by the three-dimensional blocks. The cork-covered prisms can be stacked to form a complex that suits the needs of each customer. In 2017, the architect-designer David Ajasa-Adekunle pushed the concept further by creating prefab hotel complexes using the same principles in a project called *Tetra Hotel.*

CORK in Design

/ Teapots, "The Whistler"
© Amorim collection, designed by Raquel Castro

/ Cork floor covering, Cork My Way collection
© JPS Cork Group

Intermittent paving with convex polygons
© SusanSmart–Photography courtesy of the London Design Festival

From Cork to Museum / FAT (Fashion Architecture Taste) Architecture

/ Installation at the Victoria and Albert Museum
/ London, England, 2013
/ www.fashionarchitecturetaste.com
/ www.modernarchitect.co.uk
/ www.samjacob.com

From the Renaissance to modern times, there has been but one step ... taken by the FAT Architecture firm in 2013 during the International Festival of Design, London Design Week. In collaboration with the Amorim company, the architects created a trompe-l'œil floor. This geometric paving combines various hues of dyed cork to create the floor. It was a kinetic project, located at the thresholds of design, architecture, and art installation. It evokes the evolution of the individual and the widening of experience.

The ambiance is muted thanks to the cork tiles. C-Tile from pomm Design is used on the ceiling. Cork stools complete the harmony of materials.
© David Foessel

A Cocktail of Cork / **CUT Architectures, pomm Design**

/ Le Syndicat cocktail bar
/ Paris, 2014
/ www.cut-architectures.com
/ www.pomm.pt

The art of cocktail mixing is about the subtle dosing of refined ingredients. To create a space that speaks to its neighborhood, architects and designers played with various associations in order to tell a story and arrive at the concept of the speakeasy. Traces of the space's previous incarnations were preserved and integrated into the project, like this *sol mosaico* from the old 1960s bistro with its hints of hidden gambling rooms. Quilted gold curtains, the curved countertop, and other precious details create a counterweight to the severity of the raw surfaces. The juxtaposition of the two creates the mysterious ambiance of a noir film.

Sofa Brick is made in Japan by the Nagayanagi company: the pieces can be spread out throughout already existing areas, like space invaders, thus transforming any surface into a comfortable support. This feeling of comfort is heightened by the pillow-like form created by the shape and soft touch of the cork.

Parasite Furniture / Aya Koike

/ Sofa Brick, paving module
/ Edition: Cork+, www.corkplus.com
/ Japan, Indonesia, since 2005
/ www.ayakoike.com
/ www.nagayanagi.co.jp

At first, this modular concept was inspired by a reflection on the modalities of urban life: we erect walls to limit the thresholds of private space, and so they are representative of the social distance that separates us from each other. In the tumult of daily life, our streets don't always offer shelter to pedestrians. How can we erase limits to make public spaces more welcoming? Brick, a common and standardized unit, is an enclosure that creates a limit. Deconstructing a wall to then transform it into a bench is what Sofa Brick offers. The final product developed for interior and exterior design inherited this concept of *parasite design*.

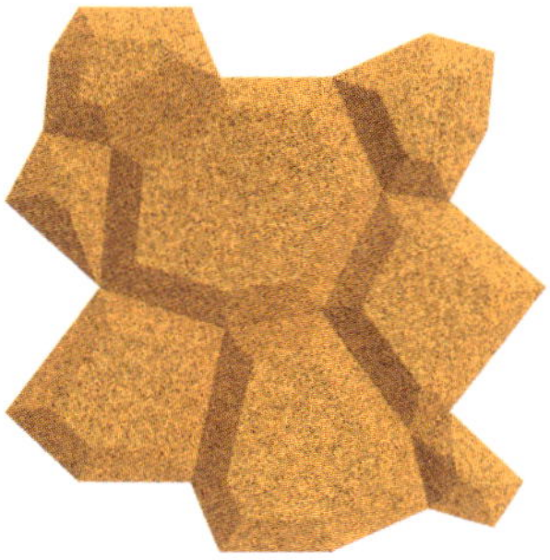

Organic Blocks sculpted wall coverings are dyed using an innovative immersion procedure. They are manufactured using molds, which gives them a homogenous appearance and enables their production in industrial series.

Organicity / Muratto by Carlstahl

/ Organic Blocks, sculpted wall covering
/ International company, award-winning in 2015 and 2016
/ www.muratto.com
/ www.carlstahl.fr

Wallpaper has become fashionable again through the use of colored patterns and modern shapes. Sold in paving strips or units, these design elements can be glued like a sticker in a specific space to create wall decor. The advantage of cork combined with these sculpted shapes is sound absorption: according to the manufacturer, the sound absorption of a wall covered in these blocks is improved by up to 30%. This means that these new wall decors offer decorators and architects joyful possibilities for surfaces while improving the comfort of the living space.

1 / 2 / Cork wallpaper; painted cork wallpaper printed with contemporary or classic designs

3/ The "Sundraw" shutter designed by Joana Fraga and Azambuja Varela

A Universe of Cork / Pearl Cork

/ Exclusive cork furnishings
/ Portuguese company, award-winning designs in 2015 and 2016
/ www.pearlcorkdesign.com

The Pearl Cork brand has declared that it will focus on cork. It offers elegant and refined designs. Certain objects like the "Kuku's Nest" cradle or the NuSpa collection are sculpted from solid cork using a five-axe milling machine. The "Segment" folding screen uses a composite white cork unit as a hinge, allowing it to move like a wave. The use of cork in its paper-like form is perfectly highlighted through various wallpaper collections and can even be used as a window blind, as seen in the customizable slatted shutter, "Sundraw."

2

3

4

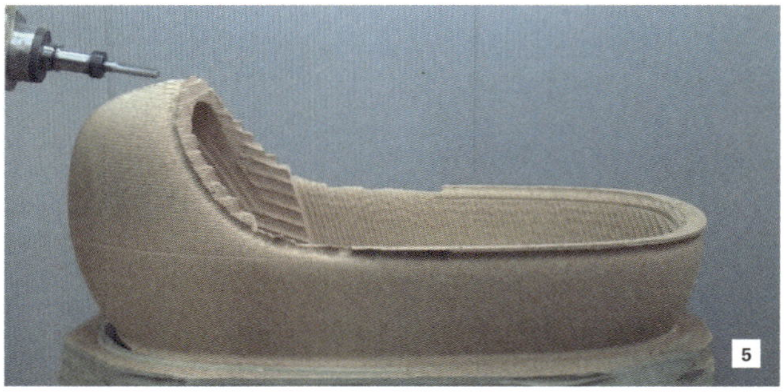
5

6

7

4 / 5 / Produced using digital milling of a large solid block

6 / 7 / "Kuku's Nest" cradle and the "Segment" folding screen were conceived by the designers Joana Fraga and Azambuja Varela.
© Pearl Cork

8 / 9 / 10 / NuSpa collection

8

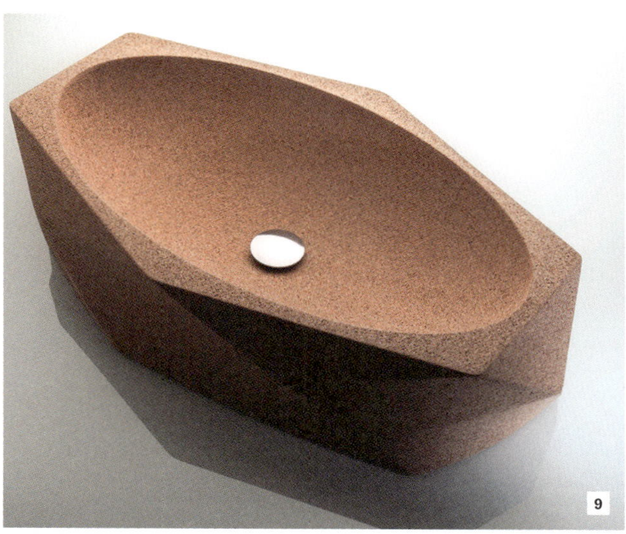

9

10

The "Float" lamps designed by Benjamin Hubert. Made by hand using a wooden lathe, these bell-shaped objects are massive and beautiful. The lamp is made by slowly removing material from the large block until the final shape is achieved.

A Lamp That Floats / Benjamin Hubert/ Studio Layer

/ "Float" lamp, suspension
/ Design and fabrication: Layer
/ London, 2013
/ www.layerdesign.com

"Float" is a series of lamps made entirely out of cork. When they are illuminated, their surfaces emit a warm light. The cork used is recycled from the Portuguese wine cork industry. Using traditional woodworking techniques, the blocks of wood are sculpted with a lathe to create massive lampshades. The production process generates a significant amount of waste, but this is recycled to make new blocks that can be made into new lamps.

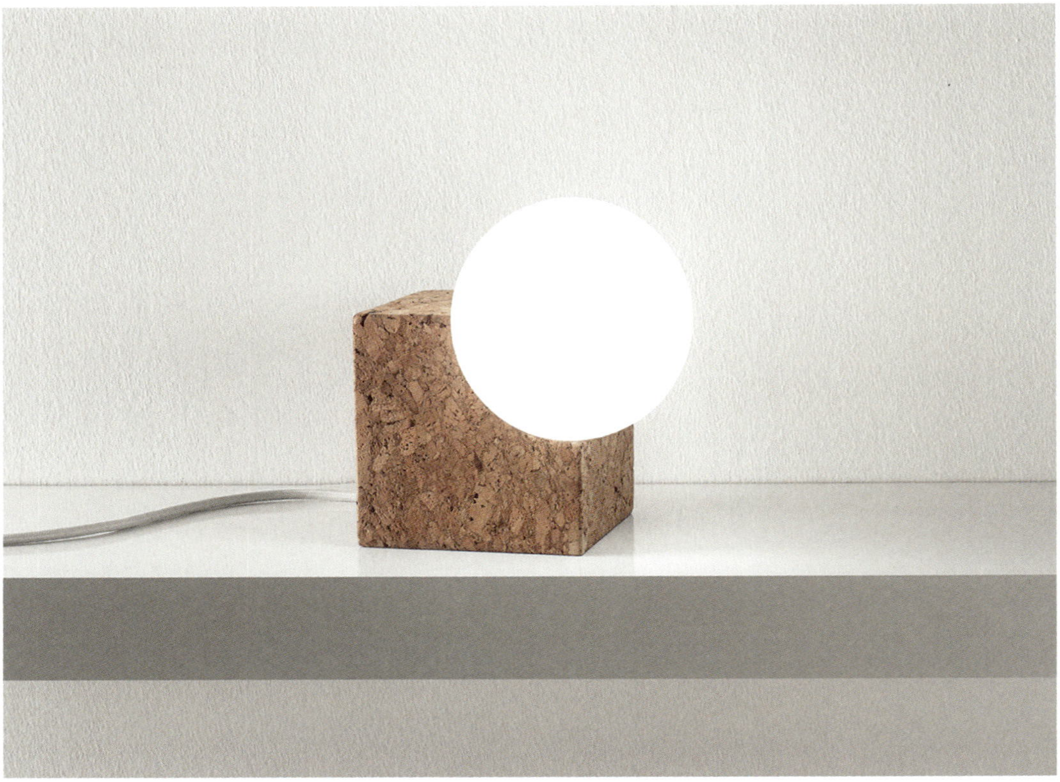

In the book, *The Hidden Dimension*,[1] the anthropologist Edward T. Hall brilliantly demonstrates the impact of bad acoustics on group relationships, for both animals and humans, as can be seen in a typical company situation. Choosing furnishing accessories manufactured in absorbent materials like cork can contribute to correcting the ambiance of a space.

1 *The Hidden Dimension* is an essay published in 1966 written by the American anthropologist Edward T. Hall.

The Eye Listens / Miguel Arruda

/ "Acoustic Lighting" (lamps), "Spherical" (recliner)
/ Editors: Exporlux ("Acoustic Lighting"), Mochevo ("Spherical")
/ Portugal, 2014
/ www.miguelarruda.com
/ www.exporlux.pt
/ www.movecho.pt

Architects are very interested in the relationship between the body and the space it occupies. Miguel Arruda has worked with anthropomorphic sculptures since the 1970s. The changes of scale brought him to develop cork furniture inspired by the human body. The spherical recliner takes this work even further. The form of the object evokes the body's curves, and the material is soft, like skin. The particularity of these stream-lined acoustic lamps is that they improve the acoustics of the spaces in which they are hung. "The eye listens ..."

The “Spherical” recliner with its simple acorn-like shape has received several international prizes.

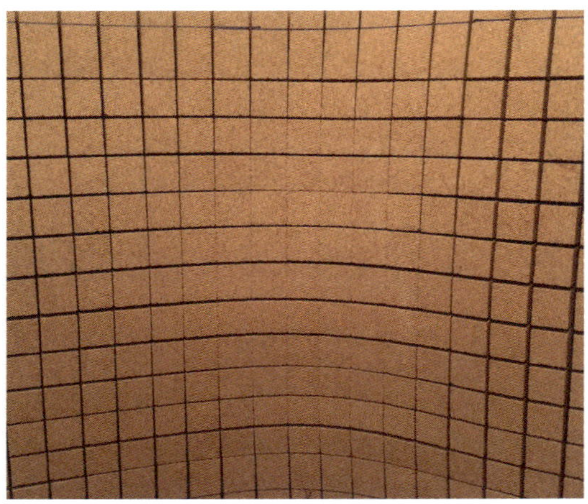

The "Cortiga" sun lounger is made by hand in a workshop. During our interview, Daniel explained that the work of layered, alternate cuts to create a hinge shape takes about an hour. Comparatively, the same shape cut using a digital process would require four.

Living Hinges / Daniel Michalik Furniture Design

/ "Cortiga" (sun lounger), "Sway Stools," "Rook Stools"
/ Brooklyn, New York, USA, 2007
/www.danielmichalik.com

The work of a designer is intimately connected to the material. If some designers move away from formal abstraction, others go deeper with repetition and observation of the same materials to perfect their technique and, in this way, orient their craft into a direct expression of their dialogue with the raw material. Daniel Michalik, who began experimenting with cork during his studies, still considers the potential of the material as "the original language of design." He creates joyful objects that are sculpted into lively shapes from the raw cork mass.

Two stools: "Sway Stools" and "Rock Stools." The first stools are crafted using delineated layers whose cutouts create a spring effect. The second set hints at a mountain with hidden treasures. These pieces are sold exclusively at the Kinder Modern Gallery in New York.
www.kindermodern.com
© Daniel Michalik

1

2

3

4

1 / 2 / "Topography" is a concept that represents certain regions of Portugal where cork is grown.

3 / "Lagarta" is a modular seat with endless combinations.

4 / "Corqui" is a cork recliner made from solid cork, either black or white.

5 / 6 / 7 / "Vinco" is a foldable chair that can be assembled without glue or screws. It is composed of two layers of white composite cork for the seated section and a rubber-cork agglomerate for the back as a hinge. The recessed areas make it stackable onto a tube-like structure that works as a foot.

A Design Brand / Corque Design

/ Designers: Toni Grilo ("Vinco" cork chair), Ana Mestre ("Lagarta"), Luis Pessanha ("Topography"), Pedro Silva Dias ("Corqui")
/ Portugal, 2009
/ www.corquedesign.com

This evocatively named brand developed out of the collaborative work of a team of young Portuguese designers working with cork, which is one of Portugal's national resources and also one of the world's most environmentally efficient materials. The team experimented with different techniques and shapes to arrive at a shared collection whose common theme is its translation into a small series of contemporary objects. The brand's approach has spread through distribution and research and with its participation in numerous design exhibitions throughout the world.

5

6

7

Thanks to the wise selection of the designer, Blackcork met its challenge and now offers interior and exterior furnishings in black cork. Black cork is rot-proof and warm, and much is used in designing contemporary gardens. The use of raw cork with streamlined shapes is the leitmotif of this collection. The projects presented are designed by Toni Grilo.

1 / O stool

2 / Omega collection

3 / Tosco lamp

4 / Cut collection

5 / Boolean lamp globe

6 / Ypsilon collection

7 / Layer coffee table

Cork Line / Blackcork

/ Editor: Blackcork—Sofalca
/ Designer: Toni Grilo
/ Portugal, 2014
/ www.blackcork.pt

The Sofalca company has been creating products manufactured from the processing of black cork since 1966. Nearly fifty years later, the brand entered the contemporary small furnishings market. To do this, it hired Toni Grilo as artistic director to help create its line. And in his wake, the company's young and talented designers collaborated to produce objects that honored this expanded material. The cork granules are infused with water vapor and then pressed. The suberin, a natural resin contained in cork, agglomerates the components to form blocks that can then be worked. Using this manufacturing technique, it is possible to imagine designs that are both simple and expressive. This is Blackcork's flagship line.

2

3

4

5

6

7

The chair's components were produced in the workshops of Carlos Ortega. "Corkigami" is created in two versions, beige and brown, on an oak base.
© Carlos Ortega

Origami / Carlos Ortega

/ "Corkigami," chair
/ Spain, 2013
/ www.carlosortegadesign.net

Creating a chair entirely out of cork seemed like an impossible challenge. But a series of experiments to transform flat planks into a three-dimensional shape ultimately resulted in a curved shape. "Corkigami" looks like a folded leaf and is reminiscent of work by Paul Jackson, whom we presented in our earlier book, *Un nouvel art du pli*.[1] For this chair, the designer and manufacturer used a crafted procedure of laminating to fine-tune a specific material. The shape of the cutout pattern permitted a curved fold but was sufficiently rigid to enable a soft and comfortable seat.

1 *Un nouvel art du pli*, éditions Alternatives, 2015.

1

2

3

Production stages:

1 / Applying the varnish to the prelaminated cork

2 / 3 / A cutout of the pattern

4 / Gluing and shaping the chair while it is pressed in a template

5 / Piercing and finishing

4

5

Usually the cork is dyed by adding pigments.
The thirteen hues, ranging from dark brown to light brown, are obtained by heating the cork for various lengths of time.
© Pedro Sadio and Maria Rita
© Amorim

In Space / AL_A, Amanda Levete

/ Cork Kit
/ Made for the Metamorphosis project by Experimentadesign, www.experimentadesign.pt, in partnership with Amorim, www.amorim.com
/ London, 2013
/ www.ala.uk.com

Thanks to advances in parametric modeling, the geometries of traditional paving now offer fascinating possibilities in terms of combinatory geometries. Within the framework of an experimental cork project, Amanda Levete, the architect and designer, offers two modular objects that can be stacked to create a new kind of shelving. The various layers of differently colored cork, once sanded, create an extraordinary motif.

"Pushpin" is created from solid cork and can be used, depending on how its placed, as a chair or as a low table. The object honors the traditional corkboards so loved by creative types. This homothetic object is a great example of the imagination and humor of the Taiwanese designer. Kenyon Yeh is the director and founder of the publisher Editions Esaila. The brand calls upon designers from around the world.

Double Face / Kenyon Yeh Design Studio

/ "Pushpin" stool and side table
/ Editor: Kenyon Yeh—kenyonyeh.com; Esaila, www.esaila.com
/ Taiwan, 2012

Inspired by the recurring theme of pins in today's world, the designer decided to pin a simple, evocative piece of furniture to its collection. The piece is special in that it can be used as a stool or as a small table to hold one's coffee. Cork, natural or dark-tinted, adds a comfortable and warm dimension to this multifunctional piece of furniture. The simplicity of the shape highlights the material and the objects associated with it, but without them taking over visually. It's true that much can be done with very little.

The "U.M.O." is made of four plates of full cork cut with a digital milling machine before being glued together. They are then lightly sanded. The tubular insert is made from thermo-lacquered aluminum.

When sculpting from a piece of solid cork, it is easy to create a small table or a bench that is both solid and versatile.

Unusual Usual / UM Project

/ "U.M.O.," bench and low table
/ Brooklyn, New York, USA, 2010
/ www.umproject.com

With the introduction of digitally programmed machines, many design studios have seen their practices evolve. By associating the virtues of hand-worked objects with these machines, small series of objects are made possible. UM Project, an independent and collaborative studio, uses functional and multicolored shapes to create unusual products. The founder, François Chambard, believes in "the ideal of the artist designer of the twentieth century." A simple process using "honest" materials led to the creation of this particular example of the designer's alien-like polymorphic furniture.

The designer began with a round-edged quadrilateral, a shape that is in line with the fashion of producing an object by removing material ...
© Amorim

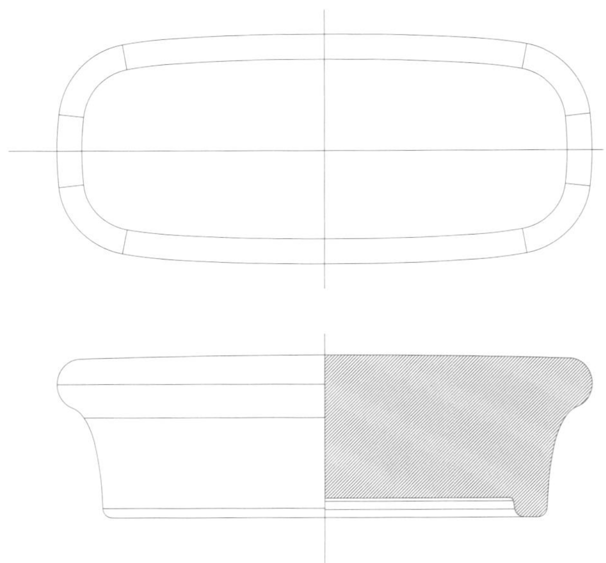

Sculpted from a Solid / Naoto Fukusawa

/ Cork Bench
/ Made for the Metamorphosis project by Experimentadesign, www.experimentadesign.pt, in partnership with Amorim, www.amorim.com
/ Tokyo, Japan and Milan, Italy, 2013
/ www.naotofukasawa.com

Every project includes some element of the unsaid. Japanese designer Naoto Fukusawa has named this attitude "Without Thought," and it is the source of simple-seeming designs that, nevertheless, touch upon a shared unconscious element. The bench sculpted from solid black cork evokes a block of stone with some hint of a bottle cork. The result is a bench that is simple, yet endowed with a certain Japanese formalism. Cork is a hydrophobic material with a temperature close to that of the human body. Because of its insulating properties, it provides, even on the outside, an agreeable tactile experience.

S
180
18/7
22/7
2ª

180
S
S
180
18/7
22/7
2ª
2ª

For this medicinal herb pot, the cork was sculpted from a solid piece and set inside a bamboo tripod connected with a traditional technique that uses hemp ropes. Cork—a material that is light, that breathes, that is waterproof—is an ideal vessel for carrying a living organism.

Melting-Pot / Toriga Happi

/ Cork planter pot
/ Team Millie Schwier and Nozomi Nakabayashi
/ London, England, 2016
/ www.theshowroom.org
/ www.torigahappi.com

Collaboration between individuals is the key to a new form of relocalized industry that is by the people and for the people. This is what London's The Showroom gallery offers through experimentation and collaborative programs for creative individuals of all backgrounds. The idea is to create a crisscrossing of various fields in order to produce works that reflect contemporary trends in terms of knowledge sharing. This itinerant medicinal herb pot is the result of a collaboration between a sculptor and an architect. The goal was to create something transportable to different community sites—a structure that questions the border between the state of nature and the urban spaces of today.

"Cork Macaron" is made of a cork sheath wrapped around a nontoxic urethane mousse to make a low-cost cushion that is available in three colors.
© Aya Koike
Yosuke Shibata

Tender Age / Aya Koike

/ "Cork Macaron" (cushion), "Tumiki+" (building blocks)
/ Edition: Cork+, www.corkplus.com
/ Manufacturer: Nagayanagi, www.nagayanagi.co.jp
/ Japan, Indonesia, 2013
/ www.ayakoike.com

Small cushions in simple shapes in an assortment of attractive colors—this is what the Japanese designer Aya Koike has created for children. This might seem simple, but the project required numerous manufacturing trials before the final product was achieved. Children love it, as they do the intellectually stimulating game of building blocks made of waterproof, floating cork. These block can form a composition of compatible shapes with natural colors and feeling. The pieces are slightly soft, with rounded edges for discovering or rediscovering the joys of architecture. Cork is a healthy material: nontoxic and hypoallergenic, it is perfectly suited for children.

The "Tumiki+" building blocks game is available in two or three sets of twenty-five or thirty-five colors. The buyer can select a cork-covered case as well.
© Aya Koike

1 / "Obab" is a swivel chair with an off-center spin. Designer: C. Ghion

2 / "Cocoon" is a protective space in which one can hide, read, or lie down to rest. By pushing the cork buttons, the child gains a panoramic vision of the world. Designer: Binôme © Playtime Paris

The ABCs of Art / Play with Design

/ Designers: Christian Ghion, www.christianghion.com ("Obab"); Binôme, www.binomedesign.com ("Cocoon"); Stéphanie Marin, www.smarin.net ("Play YET"); Valerie Windeck, www.valeriewindeck.com ("Big Puzzle"); Marine Peyre, www.marinepeyre.com ("The King of My Room")
/ Manufacturer: Agglolux, www.agglolux-cbl.com
/ Play with Design exhibition, 2nd edition, organized by Playtime Paris
/ Paris, New York, and Tokyo, 2014
/www.iloveplaytime.com

Selecting a child's toy is an important task. The shapes, colors, materials, textures ... all contribute to the child's cognitive development. And for a designer, play extends functionality; children's imaginations are unlimited. Giving beautiful objects to a child will contribute to the creation of positive inner images and to the development of his or her future skills. The Playtime International Salon provides the occasion for "childhood professionals" to discover new trends. The exhibition curator, Isabelle Poulet, decided that each year would highlight a different material. Talented designers were invited to collaborate with raw material manufacturers to create "The ABCs of Art" of the future.

2

3

4

5

6

3/4/5/6/
"Play YET" is a life-sized game set with which adults and children can create furniture. Materials: cork, wood, fabric, rubber
Designer: S. Marin
© Playtime Paris

7

8

9

7 / 8 / 9 / “Big Puzzle” is a giant puzzle made of cork.
Designer: V. Windeck

10 / “The King of My Room” is a clothes rack that opens and closes and onto which one can hang many items.
Designer: M. Peyre

10

pomm

Cork is an excellent substitute for plastics. It is recyclable, durable, has a satisfying "vibration" for the eye, and a green lifecycle.

Shock Absorber / pomm Design

/ "CorkCase"
/ Portugal, 2013
/ Limited series of 4000 pieces
/ www.pomm.pt

The introduction of tablets and the greater public's craze for these portable devices has changed our daily lives. They have also naturally become an area of interest for designers. Accessories to work with these tablets are also highly sought for their stylish variations and ergonomic properties.

This cork protection with its modern appeal, which was created in a small series for lovers of innovation, highlights cork's protective qualities. It is a shock absorber and scratch protector, and it's soft to the touch. From an aesthetic perspective, it creates a naturally elegant and high-tech case.

CORK in Fashion

/ Pelcor bag, Spring-Summer Collection 2016

/ Natural Waves, wallpaper

The cork umbrella made Pelcor famous. This iconic object has become a must-have. Pelcor also offers a diverse collection of cork "leather-work" in extraordinary colors and textures that highlight the possibilities that cork offers to fashion designers.

1 / 3 / Fall-Winter Collection 2014

2 / Spring-Summer Collection 2015

4 / Umbrella, Fall-Winter Collection 2014

Cork Skins / Pelcor

/ Accessory Collection, Spring-Summer 2016, Spring-Summer 2015, and Fall-Winter 2014
/ Portuguese brand
/ www.pelcor.pt

The leap from champagne cork to fashion is a big one. The family-owned Pelcor company went bold in 2003 when they changed their focus by launching their famous cork umbrella. Based on the umbrella's commercial success, Pelcor now develops a line of elegant and modern fashion accessories. Cork, a high-quality renewable product, enables the company to make creative and multicolored shapes with a wide range of textures. The fabric that is made, an alternative to animal leather, is appealing because of its softness and its ecological qualities.

In a 2014 collection, Repetto featured cork at the forefront of its line. Similar to leather but 100% plant-based, and both elegant and colorful, cork is a compelling choice. Wearing this soft and comfortable material is a bit like having an oak tree on your feet—like the tree, you have an elegant posture and deep roots!
© Repetto

Enter the Dance / Repetto

/ Arty Cork collection
/ Paris, 2014
/ www.repetto.fr

The universe of Repetto conjures up the airy elegance of classical ballet dancers. The brand offers articles for dance and "leatherwork" accessories, but also an entire similarly inspired wardrobe. In 2014, it introduced novel pieces in cork, a material that was relatively unknown to the general public. The Arty Cork collection features cork in a range of multicolored composite designs. The slight and delicate accessories created for today's woman present an enthusiastic contrast to classical notions of cork. With Repetto, cork enters the dance: a revelation.

repetto

MLH is a brand of accessories by Michelle Lowe-Holder, known for her sewing and complicated folding techniques. Each collection is conceptualized and conceived in her London studio before each handmade piece is produced.

Adorn Oneself with Bark / Michelle Lowe-Holder

/ Elisabeth jewelry collection, 2013
/ London, England
/ www.lowe-holder.com

Michelle Lowe-Holder appreciates cork for its unique texture, for its "soft as butter" feeling… She uses it systematically as a liner for her handbags. Packaged in rolls and feeling much like leather, the fabric can be cut, shaped, and sewn. In 2013, Lowe-Holder highlighted her favorite material in the Elizabeth jewelry collection, which was done in a medieval fantasy style. The collection highlighted cork in its necklaces, bracelets, and other jewelry designs. Manufactured in Italy, the natural dyeing techniques create a highly graphic look of alternating strips. The conception stage occurs in her London studio, where the creator prefers a direct approach: without any sketches, she confronts her materials, relies on her intuition, and allows a design to take shape.

Elected Designer for Tomorrow 2011, the talented German designer Lena Hasibether creates fashion accessories and sleek-lined clothing out of cork. Lena combines the rich texture of this material with cuts that are suited to modern life, thus inventing a very particular elegance. Cork creates a high-quality and original fabric that draws attention and curiosity.
Model: Monika Cieslar
© Björn Hokamp

Simply Cork / Lena Hasibether

/ Collection AW11, clothing and accessories, 2011
/ Bielefeld, Allemagne
/ www.lenahasibether.de

The natural origins of cork, a recyclable resource, is the main theme of Lena Hasibether's Fall-Winter 2011 collection. It represents the characteristically sleek style of all her work—a harmonious collage of light and fluid elements broken up by geometric shapes. A subtle combination of silk, viscose, and cotton are mingled here with cork whose delicate texture is printed directly onto the cotton. Hasibether's fashion accessories are a testament to her attention to detail and, in particular, to the textures made possible using cork fabrics.

At the Gates of Perception

Suzaan Heyns

/ Collection presented at Mercedes-Benz Fashion Week, 2013
/ Johannesburg, South Africa
/ www.suzaanheyns.com

Inspired by the sculpted doors of Stone Town in Zanzibar, fashion designer Suzaan Heyns brought cork into her 2013 collection that was presented during Mercedes-Benz Fashion Week in Johannesburg. This "plant leather" has exceptional qualities: it is strong, waterproof, and soft to touch. Wearing it is a little like dressing in the skin of a tree. Composed of successive layers, like the laces used for decoration in architecture, these magnificent pieces of clothing are an ode to perception. Each layer reveals further mysteries to be discovered …

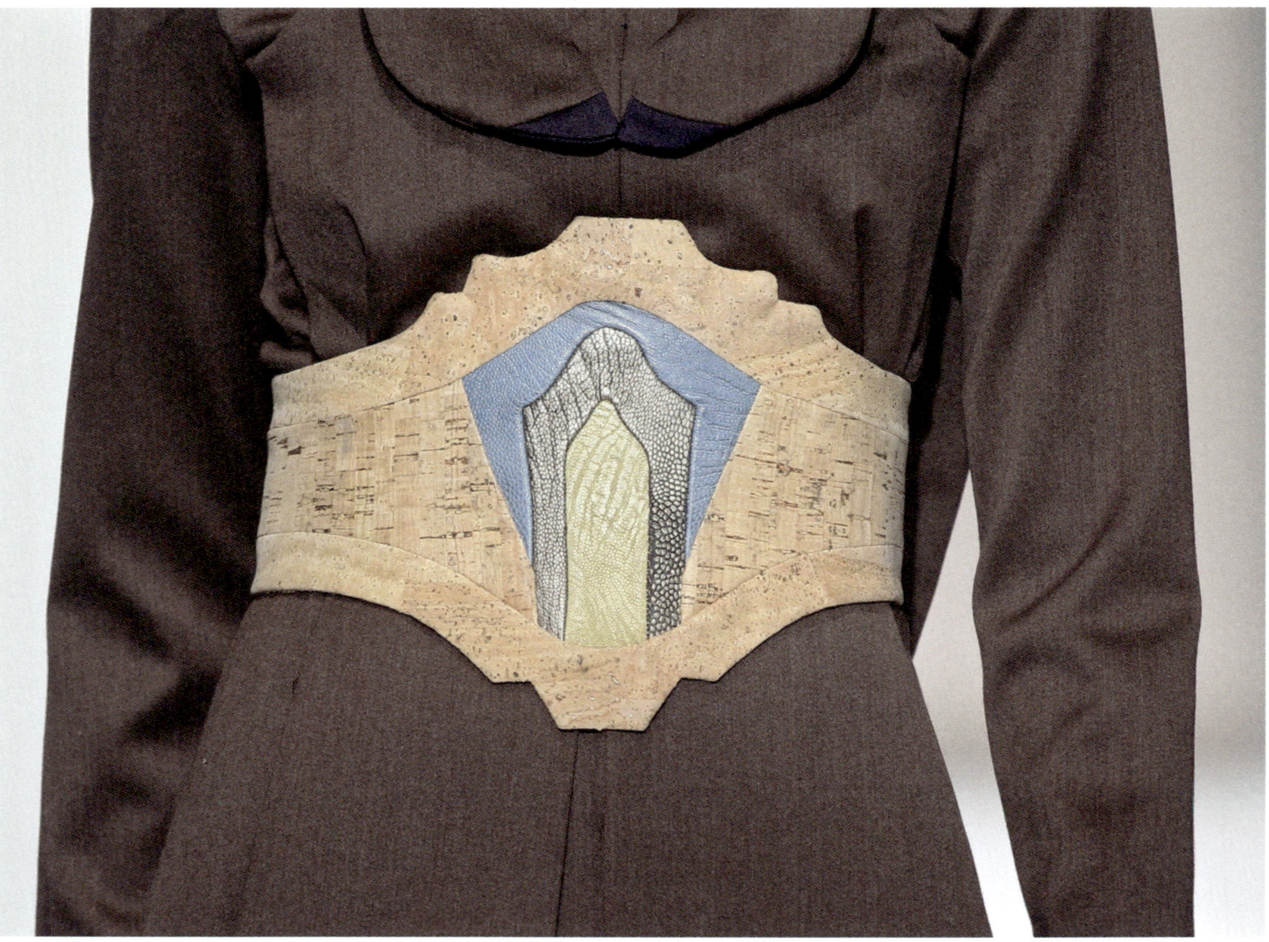

CORK in Art

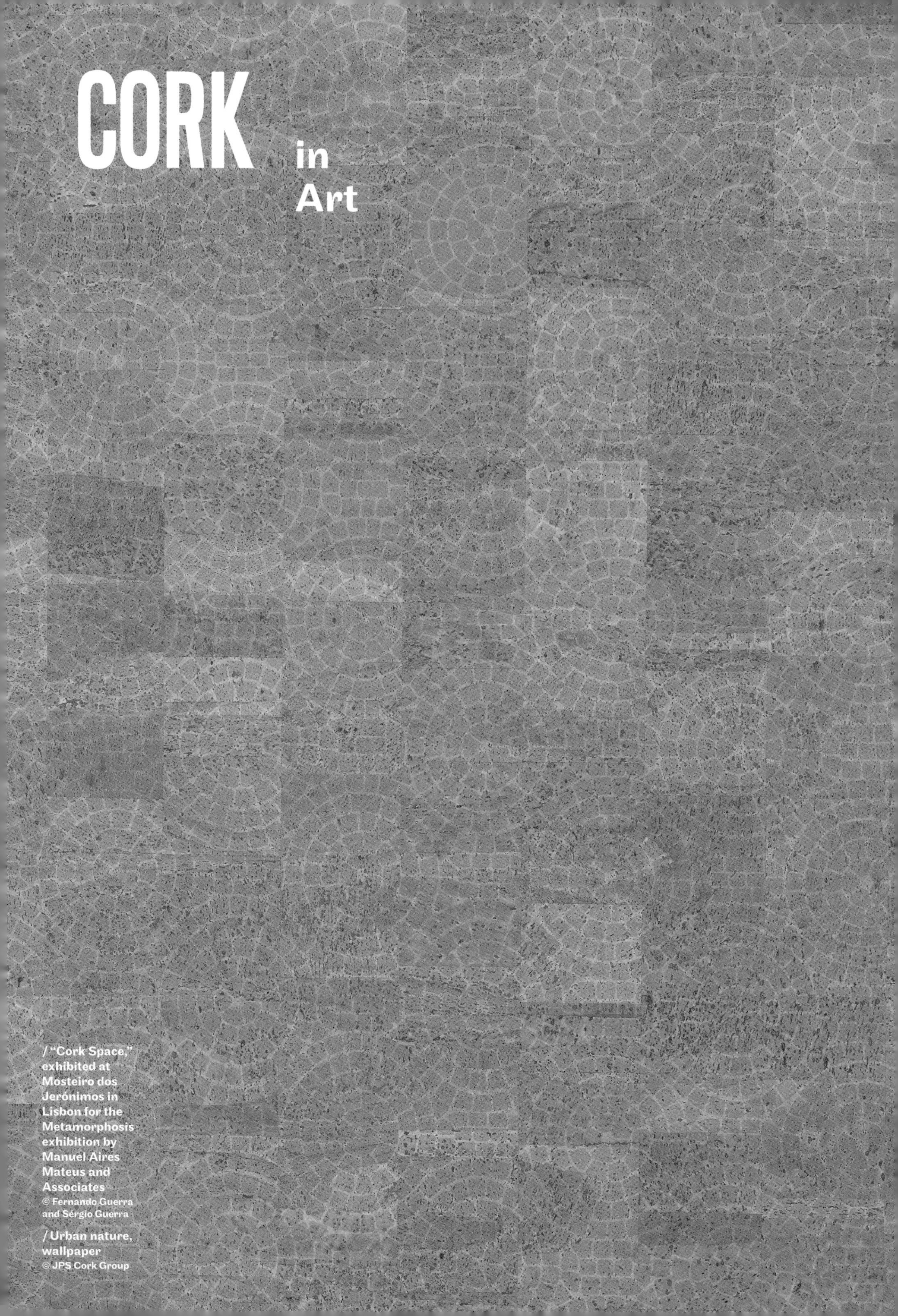

/ "Cork Space," exhibited at Mosteiro dos Jerónimos in Lisbon for the Metamorphosis exhibition by Manuel Aires Mateus and Associates
© Fernando Guerra and Sérgio Guerra

/ Urban nature, wallpaper
© JPS Cork Group

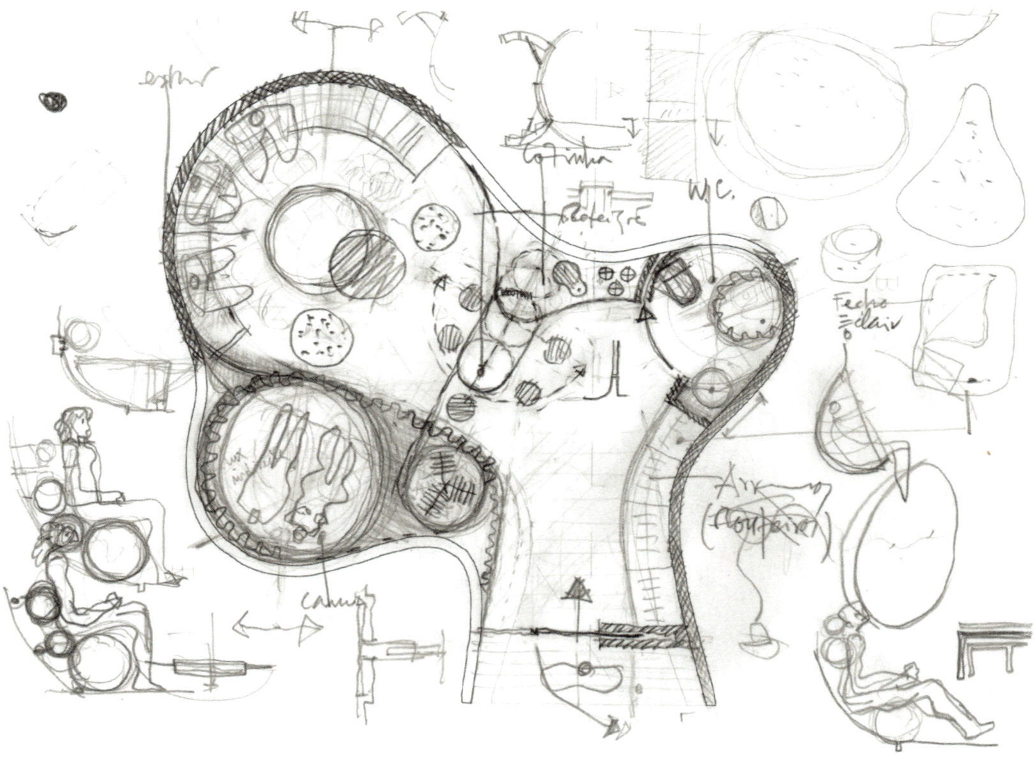

This sculpture, the work of an architect, transports visitors into a curved space without the usual points of reference made by corners and right angles. Cork, a material that is easy to shape, creates a double curve. It gives the space a muted feel —black on the outside and white on the inside, like bark—and it is silky to the touch, like skin.
© FG+SG–Fernando Guerra

The Heart of the Work / Miguel Arruda

/ Habitable sculpture
/ Lisbon, 2010
/ www.miguelarruda.com

Artist and designer Miguel Arruda is working with scale in this piece. Entering the soul of a work is like changing dimensions, and this is precisely what Arruda did for this monumental sculpture exhibited in the garden of the Belém Cultural Center in Lisbon. Based on an earlier work done in metal, which was 3D scanned, the artist designed this Lewis Carrol–esque homothety. Take a moment to imagine your sensory perceptions from inside a tiny object: see how the combination of cork and curves sets the stage for a journey inside the womb.

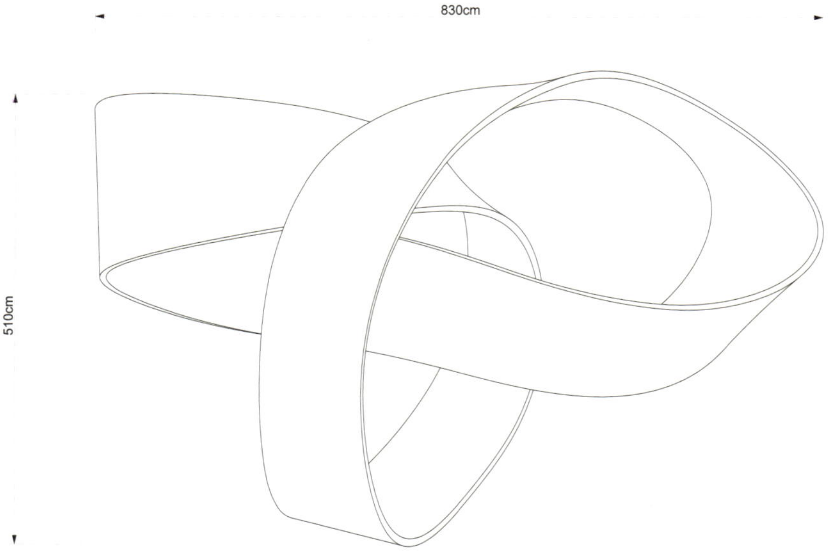

Materials were selected for their recyclability and organic sourcing. Linen fibers, nonsaturated organic resin from soy, and an Amorim cork heart were used to make this work. Phase changing bearings integrated into the resin regulate the surface temperature to make it comfortable to the touch at any time of day. The final layer was a self-cleaning finish made from nanoparticles of active titanium dioxide (TiO_2) and nanostructured silica (SiO_2).

Finally, the kinetic energy generated by visitors was converted using a piezoelectric system and transferred to an interactive low-consumption lighting system. Flexible solar panels were also used on the curved shape.

Biomimetic / 3XN Architects

/ "Learning from Nature"
/ Louisiana Museum of Modern Art, www.louisiana.dk
/ Denmark, 2013
/ www.3xn.com

For this nature-inspired sculpture installation, the architects played with a variety of registers. The continuous figure-eight Moebius strip shape is a nod to the eternally renewing cycle of life. All components came from recent technologies and were fine-tuned to preserve the environment. The project was completed in partnership with both engineering and industrial offices.

1

"The exterior walls of the pavilion have the same texture as the exterior of the cork tree. They were made using surplus materials from bottle cork manufacturing. This is the hardest part of the bark, and the one we are least used to seeing. The goal of this piece was to highlight this texture, showcasing it as something natural. Then one of the walls was covered in gold leaf so it would react differently in the sunlight."—
Manuel Aires Mateus

2

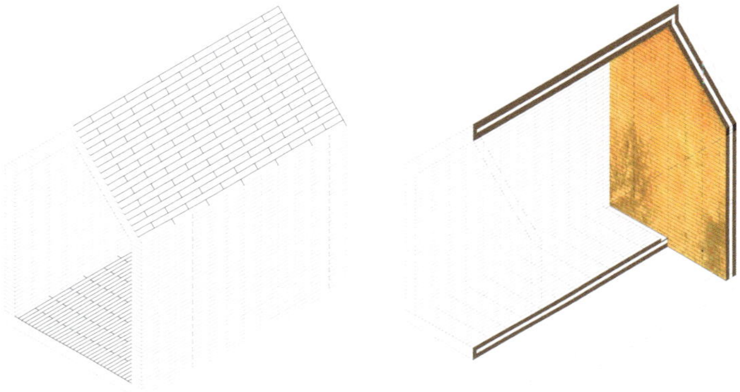

3

Metamorphosis ⁄ Manuel Aires Mateus and Associates

/ "Cork Space," exhibited at Mosteiro dos Jerónimos, in Lisbon
/ Context: created for the Metamorphosis project by Experimentadesign, in partnership with Corticeira Amorim
/ Portugal, 2013
/ www.airesmateus.com

The architecture project for this pavilion used in the closing of the Metamorphosis exhibition was to highlight an unknown aspect of cork. The designer used bark strips—waste from bottle cork production—to make the panels. This small house was like the bark of a tree: rough and vitally important.

4

Today, graffiti has moved beyond the secrecy of an underground movement, and in this graffiti work, cork puts itself on display. This piece has appeared in a company entryway and in the street, during a festival celebrating Portuguese culture.
Art by Vhils is based upon the deconstruction of the material to reveal faces that appear like the souls of the spaces they inhabit. Cork is a crumbly material that is easy to mill using digital milling techniques or to cut with hand tools. The different layers achieved with these techniques form three-dimensional landscapes that add a level of additional texture to these new-age bas-reliefs.

Multiple Faces / Vhils

/ "Contrast," bas-relief
/ PT Data Center at Covilhã, Portugal, 2013
/ Imaginarius International Festival of Street Theatre, Santa Maria da Feira, Portugal, 2013
/ www.vhils.com

Vhils—alias of Alexandre Farto—is an internationally-recognized street artist known for his wall "explosion" portraits. Today, he is commissioned both publicly and privately for monumental works. His bas-relief cork frescoes express a contemporary use of this material cherished by the Portuguese culture it comes from. The faces and shapes that Vhils designs capture a little bit of the soul of the people who wander the streets of our cities and the highways of the internet.

ISAO

VISION

Saimir Strati is an Albanian mosaic artist who was born in Tirana in 1966. Traditional mosaicists were modest artists who were mostly unknown. Saimir has created numerous performances using common objects like nails or corks and has earned his entry into the *Guinness Book of World Records*.

In this way, he has turned an outdated art form into a contemporary performance. His work asks questions about the relationship between contemporary humanity and the place of the artist in public spaces.

Bottles ... to the Sea / Saimir Strati

/ Sheraton Tirana Hotel, Albania, 2008
/ https://fr-fr.facebook.com/saimirstratii

Saimir Strati began his career as a mosaicist by restoring archeological remains in Albania. He then began creating immense mosaics using unlikely objects—nails, screws, toothpicks, scissors. During a dinnertime discussion with friends about the common denominators of Mediterranean culture, the response was unanimous: wine, of course! And its marvelous variety of bottle corks. This was the inspiration for a new monumental work of art. Strati put on a performance: he assembled 229,764 bottle corks side by side over the course of twenty-eight days. The various colors paid homage to the mosaic of Mediterranean cultures.

MONTES
CHATEAU

© Nacho and Maria Cartagena

Species of Spaces / Nacho Cartagena

/ "Cork," sculpture
/ Canada, public works since 2007
/ www.nachocartagena.com

"Cork," by Nacho Cartagena, is a city that seems to be moving and is devoid of people. This absence and the intrinsic movement of the piece evokes the exile of numerous Chileans who, like the artist, moved to Canada in 1976 to escape dictatorship. Richly detailed, Nacho's cork bas-reliefs are like an invitation for humans to live in nature. The strange shapes of the houses of the work are reminiscent of *Howl's Moving Castle*, the famous animation film by Hayao Miyazaki. The facade hides a secret passage, a kind of doorway that opens between two warring planets, like an escape from human madness ... the crumbly and fragile cork is somehow childlike. For the sculptor, cork is a malleable and warm medium, however, when sculpted with woodworking tools, it can be made into sharp edges or sanded into curves.

1

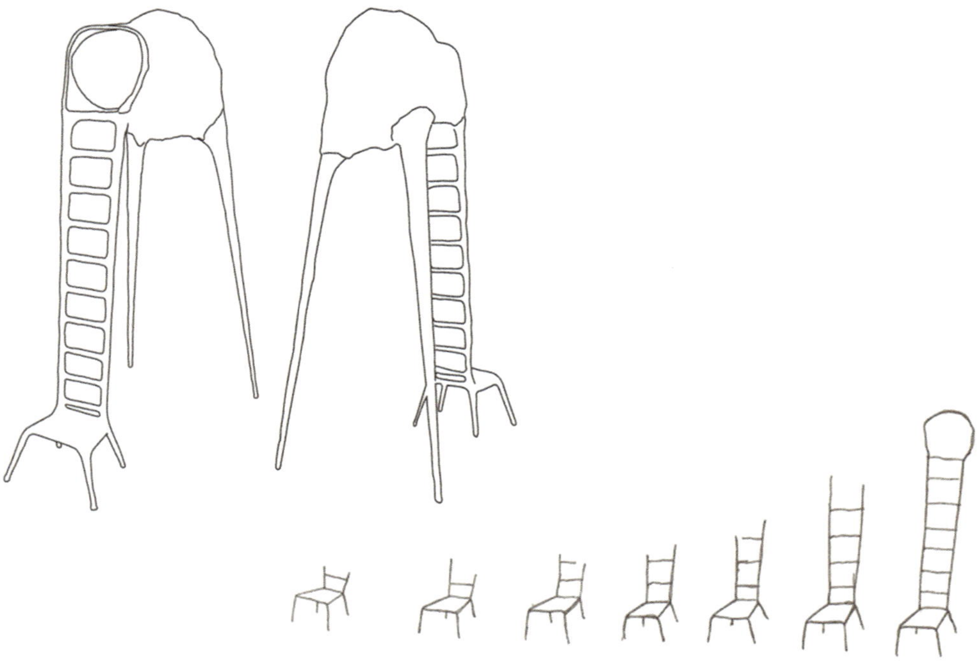

2

Chair Dreams / Nacho Carbonell, Bark Shelter

/ Sculpture
/ Groninger, Netherlands, 2014
/ www.nachocarbonell.com

By looking through the window, the chair understood that it once came from a tree. In its dreams, the chair begins to grow, like a living being, accumulating layers of bark each day, until it is a welcoming place of solace to escape the harshness of daily life. The fertile imagination of Nacho Carbonell, a Spanish artist living in Eindhoven in the Netherlands, animates his works with a lifeforce of its own. This piece from the Tree Chair Family series is exhibited at the Groninger Museum in the Netherlands. It is composed of cork granules glued to a metallic structure.

3

ANNEXES

/ Panels of Composite Cork
© Amorim

/ Urban nature, wallpaper
© JPS Cork Group

USEFUL CONTACTS

/ Associations

RETECORK European Network of Cork Production Regions:
www.retecork.org

IML Mediterranean Cork Institute:
www.institutduliege.com

APCOR Portuguese Association of Cork Industries:
www.apcor.pt

European Forest Genetic Resources Programme (EUFORGEN):
www.euforgen.org

/ Museums

Spain

Centro de Interpretación de la Sierra de San Pedro Alas, San Vicente de Alcántara:
www.sanvicentedealcantara.es

Centro de Interpretación del Alcornocal, Muelas del Pan:
www.alcornocal.com

Centro de Interpretación Etnográfica de Los Alcornocales, Alcalá de los Gazules:
www.museodealcala.wordpress.com

Centro de Interpretación P.N. Sierra del Espadán, Eslida:
www.habitatge.gva.es

Museo de Identidad del Corcho, San Vicente de Alcántara:
www.sanvicentedealcantara.es

Museu del Suro, Palafrugell:
www.museudelsuro.cat

France

Musée du Liège, Maureillas les Illas:
www.maureillas.fr

Musée du Liège et du Bouchon, Mézin:
www.ville-mezin.fr

Italy

Museo Storico delle Macchine del Sughero, Tempio-Pausania:
www.sardegnacultura.it

Portugal

Ecomuseo Municipal do Seixal:
www.cm-seixal.pt/ecomuseu- municipal

Museo da Cortiça da Fabrica do Ingles, Silves:
www.cm-silves.pt

Museu do Traje:
www.museu-sbras.com

Musée José Régio, Portalegre:
www.cm-portalegre.pt

/ Manufacturers of Cork-based Materials

Amorim
Portuguese multinational company
www.amorim.com

Bouchons Amorim
www.amorimfrance.fr

Wicanders, company of the Amorim group
www.wicanders.com

Muratto by Carlstahl
Portugal
www.muratto.com, www.carlstahl.fr

Agglolux
France
www.agglolux-cbl.com

Sedacor, company of the JPS Cork Group
Portugal
www.jpscorkgroup.com
www.divinacorkoleum.com

4Spaces
Switzerland
www.4spaces.ch

colorFabb
Holland
www.colorfabb.com

Friedbert Bleile Handel and Dienstleistungen
Germany
www.korkmosaik.com

Jelinek
Portugal
www.jelinek.com

L'écorce du chêne
Christophe Crublet
France
www.revetement-liege.com

/ Designers
(site not yet mentioned in previous pages)

Toni Grilo
www.tonigrilo.com

Pedro Silva Dias
www.pedrosilvadias.com

Jasper Morrison
www.jaspermorrison.com

Fernando Brizio
www.fernandobrizio.com

Bounoure & Genevaux design
www.bou-ge.com

/ Home Furnishing Companies Using Cork
(nonexhaustive list)

Materia
www.materia.amorim.com

Cork+
Japan
www.corkplus.com

Nagayanagi
Japan
www.nagayanagi.co.jp

Exporlux Acoustic Lighting
Portugal
www.exporlux.pt

Movecho
Portugal
www.movecho.pt

Corque Design
Portugal
www.corquedesign.com

Blackcork | Sofalca
Portugal
www.blackcork.pt

Esaila
Taiwan
www.esaila.com

pomm Design
Portugal
www.pomm.pt

La Chance
France
www.lachance.fr

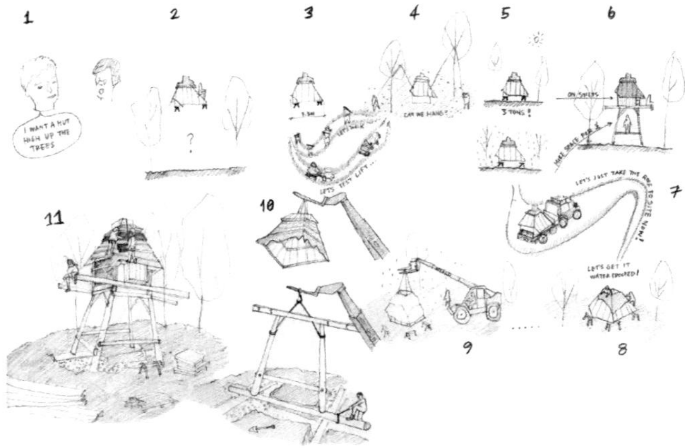

/ Sketch by Nozomi Nakabayashi for the Huts on Stilts project
Acknowledgments:
Nozomo Nakabayashi
www.nozomi-nakabayashi.com

/ "Cork Macaron" by Aya Koike
© Aya Koike

ACKNOWLEDGMENTS

We thank our families, our friends, and our guides for their support. We also thank the different individuals who contributed to this work, in particular, Sabine Bledniak and Pierre Beaucousin from Editions Alternatives, intern Thibaud Pean and stage director architect Philippe Capelier, Isabelle Poulet from Playtime Paris, and designers Daniel Michalik, Thomas Moulis, and Josette Taudou.

Special thanks to Amorim for their images and their help with our research.

/ Introduction

Joana Martins, Carlos De Jesus, Christina Amorim, Antonio Amorim, Carlos Manuel, Amorim / Experimentadesign / Eduardo Souto De Moura, architect / Jasper Morrison, designer / Mauricio O Brien, Corklab / Renaud Piazzetta, Mediterranean Cork Institute (IML) / Bernard Cave, Agglolux, Jean-Baptiste Coffin, photographer / Fernando Brizio, designer / Gonçalo Campos, designer / Raquel Castro, designer / Rui Pereira, designer / Materia par Amorim / Carlos Alberto Sá, Samuel Reis, JPS Cork Group / Michele Rondelli, Martin Reichenbach, 4Spaces / Sander Strijbos, colorFabb / gCreate / seechless / Amorim, Mercedes-Benz and Garrett McNamara, Heather Rosenstein, David Van Klaveren, Barefooters / João Luís Carrilho da Graca, architect / Friedbert Bleile, Friedbert Bleile Handel and Dienstleistungen / David Fajfer, Vera Fajferova, Carla Andrade, Jelinek Cork Group / Christophe Crublet, L'écorce du chêne / Jordi Bonet, Dr Arquitecte / Sandrine Coeur Bizot, architect

/ Architecture

Olivier Brichet, scenographer / builder and sound designer, Alexandra Epée, designer and scenographer, Flora Rich, landscape architect, International Gardens Festival, Chaumont-sur-Loire / Nozomi Nakabayashi, architect, Henrietta Williams, photographer / Rupert Maleczek, Architect Collective Archiwaste, Stefan Rutzinger, Christian Scheiber, Koge, Institut für Konstruktion und Gestaltung, Innsbruck, Austria / Gabriela Gomes, author and artistic director, Sergio Bessa, architect, João Morgado, photographer / Colin Fournier, architect, Professor Emiritus of Architecture UCL, University College London, Bartlett School of Architecture, Marysia Lewandowska, artist / Stefanie Manthey Herzog and de Meuron architectes, Ai Weiwei artist, Anya Johnson, Gallerie Serpentine, Suzanne Tóth-Pál, Iwan Baan, Iwan Baan / Antonio Moll Architect and photographer / CC, Carlos Couto, Joana Fontoura, Atelier de Arquitectura / José Carlos Cruz, Julião Pinto Leite, José Carlos Cruz Arquitecto, Sergio Guerra, FG+SG— Fotografia de arquitectura / Tiago Rocha, Branco Cavaleiro Architectes, José Campos, photographer / Nuno Oliveira, Lídia Costa, Nengenharia, Pedro Passos, Imperfect/ Vasco Magalhaes, Filipe Afonso, Arquitectos Anónimos, Ivo Canelas, photographer / Joel Esperança and Ruben Vaz, Romeu Sousa, Geral, Contaminar Arquitectos / Eduardo Carvalho, Francisco Freire, Luís Gama, Plano B Arquitectura, Sandra Pereira, photographer / David Ajasa, Adekunle, Tetra Shed

/ Design

FAT Architecture, Sam Jacob, Flick Barbur, Sam Jacob Studio, Sean Griffiths ModernArchitect, Professor of Architecture, University of Westminster, London Design Festival, Susan Smart Photography / Yann Martin, Cut Architectures, Miguel Mota, pomm Design, David Foessel, photographer / Aya Koike, Nagayanagi / Dominique David, Vasco Barros, Michael Riethmuller, Rita Pereira and Audrey Wagner, Muratto by Carlstahl / Carla Rego, Pearl Cork / Benjamin Hubert, Giselle Jonte, Studio Layer / Miguel Arruda, Mariana Vaz Antunes, Pedro Nogeira, Miguel Arruda Arquitectos Associados, Exporlux acoustic lighting, Movecho / Daniel Michalik designer, BFA product design / assistant professor, Parsons School of Design / Ana Mestre, Toni Grilo, Luis Pesanha, Pedro Silva Dias, Corque Design / Toni Grilo, Blackcork, Sofalca, photography / Carlos Ortega, design / Amanda Levette, Mary Dymond, AL_A, Pedro Sadio and Maria Rita, photographers / Kenyon Yeh, Esaila / François Chambard, UM Project / Naoto Fukasawa, Keiji Takeuchi, Naoto Fukasawa Design / Nozomi Nakabayashi, Millie Schwier, Toriga Happi, The Showroom / Aya Koike, Nagayanagi, Cork+, Yosuke Shibata, photographer / Isabelle Poulet, Martina Picaflor, Playtime Paris, Christian Ghion, design, Binome: Ingrid Michel and Frédéric Pain, design, Laboratoire Stéphanie Marin, design, Valerie Windeck, design, Marine Peyre, design / Miguel Mota, pomm Design / Pam McDonald, Royal College of Art London

/ Fashion

Ana Correia, Jay Austin, Pelcor / Stephanie Arfa, Marion Gardair, Catherine Miran, Press office, Repetto / Michelle Lowe-Holder fashion designer, Polly Penrose, photographer, Lena Hasibether, fashion designer, Björn Hokamp, photographer / Moira Jensen, Suzaan Heyns, fashion designer, Simon Deiner, SDR Photo

/ Art

Miguel Arruda, Mariana Vaz Antunes, Miguel Arruda architecture / Cecilie Østerby, 3XN architects, Adam Mørk photographer / Manuel Aires Mateus, Cassandra Carvas, Aires Mateus e associados / Alexandre Farto, Duarte Cavalinhos, Vhils Studio, Pedro Patrocínio, photographer/ Saimir Strati and mosaicart-sast, artist / Nacho Cartagena, artist, Maria Cartagena / Nacho Carbonell, artist, Paloma Castano

Despite our research, some photographers could not be identified. We ask for their understanding and will be sure to make corrections, following image authentication, in any further printings.

/ Models of the *Corkigama* chair by Carlos Ortega Design